TM 10-276

U.S. ARMY TECHNICAL MANUAL

HOT WEATHER CLOTHING AND EQUIPMENT

1970

CIVILIAN REFERENCE EDITION

UNABRIDGED HISTORIC MILITARY MANUAL ON ATTIRE AND ACCESSORIES FOR HIGH-TEMPERATURE AND JUNGLE CLIMATES

U.S. DEPARTMENT OF THE ARMY

Doublebit Press

www.doublebitpress.com
Cherry, IL, USA

Original content under the public domain; unrestricted for civilian distribution.
Originally published in 1970 by the U.S. Department of the Army.

A Part of the Doublebit Military Outdoors Skills Series.

Doublebit Press Civilian Reference Edition ISBNs
Hardcover: 978-1643892085
Paperback: 978-1643892092

TECHNICAL MANUAL
No. 10-276

HEADQUARTERS
DEPARTMENT OF THE ARMY
WASHINGTON, D.C., *15 August 1970*

HOT WEATHER CLOTHING AND EQUIPMENT

LIST OF ILLUSTRATIONS

CHAPTER 1

INTRODUCTION

Section I. GENERAL

1. Purpose and Scope

This manual provides information and guidance on hot weather clothing and equipment. It lists and describes each item, provides instructions for wear or use and fitting, and covers general cleaning and care of hot weather clothing and equipment issued to the soldier for his personal use.

2. Bases for Issue

CTA 50-901 and CTA 50-902 govern the issue of hot weather clothing and equipment to all military personnel of the Department of the Army.

3. Revisions Based on Changes in Concept and Design

The clothing and equipment described and the information given in this manual are current at the time of preparation. Changes in hot weather clothing and equipment concepts and systems and changes or modifications in design may require revisions of the material.

4. Reporting of Technical Manual Improvements

Users of this publication are encouraged to report errors and omissions and to offer recommendations for improvement. Reports should be submitted on DA Form 2028 (Recommended Changes to Publications) and forwarded direct to the Commandant, U.S. Army Quartermaster School, ATTN: ATSQM-AR-T, Fort Lee, Va. 23801.

Section II. FUNDAMENTALS OF HOT WEATHER CLOTHING DESIGN

5. Purpose of Clothing

There are three principal reasons for wearing clothing:

a. To Cover the Body. This is mainly to comply with custom.

b. To Protect the Skin. Clothing protects the skin from the elements, cuts, scratches, and insect bites.

c. To Aid Body Functions. The most critical reason for wearing clothing is to keep the body sufficiently comfortable so that the body functions normally. Under certain conditions, the best way to keep *cool* is to *put on* lightweight garments; *for example,* during intense sunlight, clothing should be worn to keep the heat away from the body. There are other instances when the best way to keep *warm* is to *take off* some clothing. *For example,* overheating causes clothing to get damp and damp clothes can be cold.

6. Basis for Clothing Design

Hot weather clothing and equipment have been designed with certain environmental conditions in mind. Nature's system for keeping man's internal organs at a nearly constant temperature is complicated and elaborate. The fact that a constant body temperature is maintained implies that the rate at which heat is produced is the same as the rate at which it is released or dissipated. The principles by which heat is dispersed from body surfaces are radiation, convection, and evaporation of water. The soldier in hot climates has many clothing and equipment problems that do not exist for the native resident. *For example,* the soldier has such problems as hygiene, body temperature control, uniform clothing, insect protection, camouflage and mission mobility.

CHAPTER 2

HOT WEATHER CLOTHING AND FOOTWEAR

Section I. CLOTHING COMPONENTS

7. General

Made of minimum weight, nonbulky material, hot weather clothing is functionally designed to protect the soldier from the sun and keep him warm at night. The outer (shield) fabric is tightly woven to provide protection from such hazards as mountain winds and insects. Yet the material is light enough to dry quickly after being soaked by perspiration or water. The clothing is loose fitting and has openings which are easily adjusted to encourage airflow for ventilation. It has adjustable snug closures at the waist, neck, midbody, and legs for protection against environmental hazards. Drainage eyelets are used when necessary.

8. Underwear

a. Description. The hot weather underwear consists of a cotton pullover-style undershirt with quarter-length sleeves and a crew neck and the drawers which are the standard thigh-length boxer-type with an elasticized waistband.

b. Use. Wear the underwear (optional) with the undershirt tucked into the top of the drawers.

9. Hot Weather Coat

a. Description. The hot weather coat (A, fig. 1) is single breasted with two slanted breast-patch, bellows-type pockets, two lower bellows-type patch pockets, and rollup sleeves. All pockets have concealed-button flaps and drainage eyelets. The front closure also has concealed buttons. The coat, which has adjustable cuffs, is made of lightweight, olive-green cotton ripstop poplin. The coat is also available in camouflage pattern (app. B).

b. Use. Wear the coat loosely outside the trousers to allow air circulation to the upper part of the body. Never wear the coat *inside* the trousers. The buttons at the front closure and pockets are concealed to prevent snagging.

10. Hot Weather Trousers

a. Description. The hot weather trousers (B, fig. 1) have a slide fastener fly, two front pockets, two hip pockets, and two bellows-type cargo pockets. All pockets have concealed-button flaps. The left cargo pocket has a smaller inside pocket, and both cargo pockets have drainage eyelets. The waist has belt loops and two waist-adjusting tabs. A tunneled draw cord is located at each bottom edge of the trouser legs. The trousers are made of lightweight, olive-green cotton ripstop poplin. The trousers are also available in camouflage pattern (app. B).

b. Use. Wear the trousers as an outer garment. Carry the special survival kit in the small inside pocket of the left cargo pocket. The other pockets are large enough to carry personal belongings as well as spare dry clothes. *However,* the pockets are not strong enough to carry ammunition regularly. Use the draw cords to tighten and blouse the bottom of the trouser legs over the boots; saturate with insecticide to prevent entry of leeches, insects, etc. Open the draw cords to permit the flow of air around the legs. This flow of air helps keep legs cool and helps to dry the trousers.

11. Hot Weather Cap

The olive-green shade 106 hot weather cap (fig. 2) with visor is baseball-cap styled. It is made of polyester and rayon gabardine cloth and has a sweatband made of the basic material.

12. Hats With Insect Nets

a. Tropical Hat With Detachable Head Net.

(1) *Description.* The olive-green tropical hat (A, fig. 3) is made of a lightweight water-repellent material. It has a full, semirigid brim, a low slope crown, an adjustable chinstrap, and an outer camouflage band with slots for in-

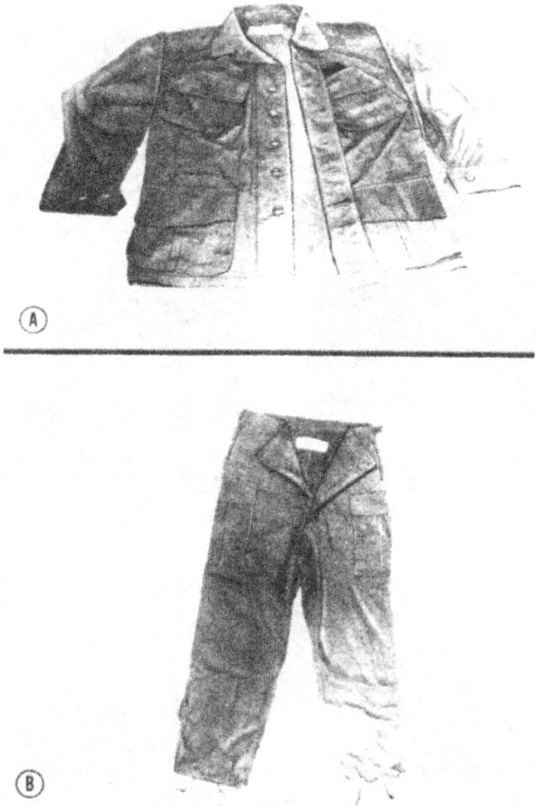

Figure 1. Hot weather coat and trousers.

serting camouflage material. The detachable head net can be carried in the pocket.

(2) *Use.* Wear the tropical hat to protect the face, neck, and head from rain and sun. The head net is easily attached with elastic loops to provide protection from insects.

b. Hat and Insect Net.

(1) *Description.* The hat and insect net (B, fig. 3) is camouflage patterned and has a cloth toppiece (no brim) with an elastic suspension that fits over the head or helmet. Metal rings hold the net away from the face and neck even when the wearer is sleeping. The bottom portion of the net has two elastic loops and an elastic draw tape.

Figure 2. Hot weather cap.

(2) *Use.* Wear the hat and insect net for protection against mosquitoes and other insects. Wear it over the head or helmet or in between the liner and the helmet. The hat and insect net, which fits over the collar in back, is held in place in front by two elastic loops that can be attached to the pocket buttons of the coat. The elastic draw tape may be drawn tight when the net is not fastened to the pocket buttons.

13. Neckerchief

a. Description. The olive-green neckerchief (fig. 4) is 36 inches by 24 inches and is made of a highly absorbent knit fabric.

b. Use. Below are some suggested uses for the

Figure 3. Hats with insect nets.

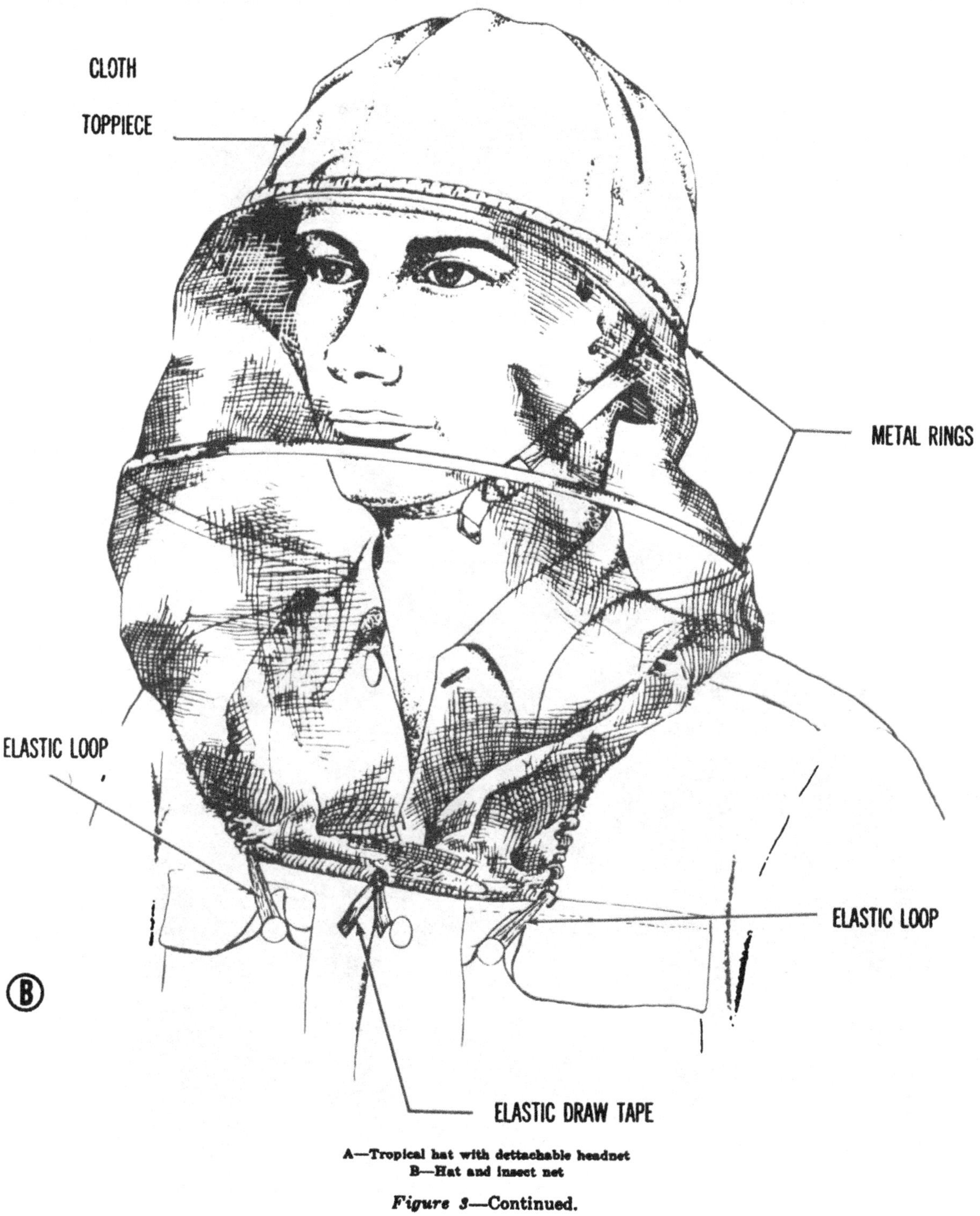

A—Tropical hat with dettachable headnet
B—Hat and insect net

Figure 3—Continued.

neckerchief. Field experience will suggest many other uses.

(1) Wiping perspiration from the face and eyes.

(2) Cleaning dirt from the hands.

(3) Cleaning ammunition before it is fired.

(4) Cleaning weapon sights.

(5) Cleaning binoculars.

14. Lightweight Poncho

a. Description. The waterproof lightweight poncho with hood (fig. 5) is made of coated nylon. It is approximately rectangular in shape with

Figure 4. Neckerchief.

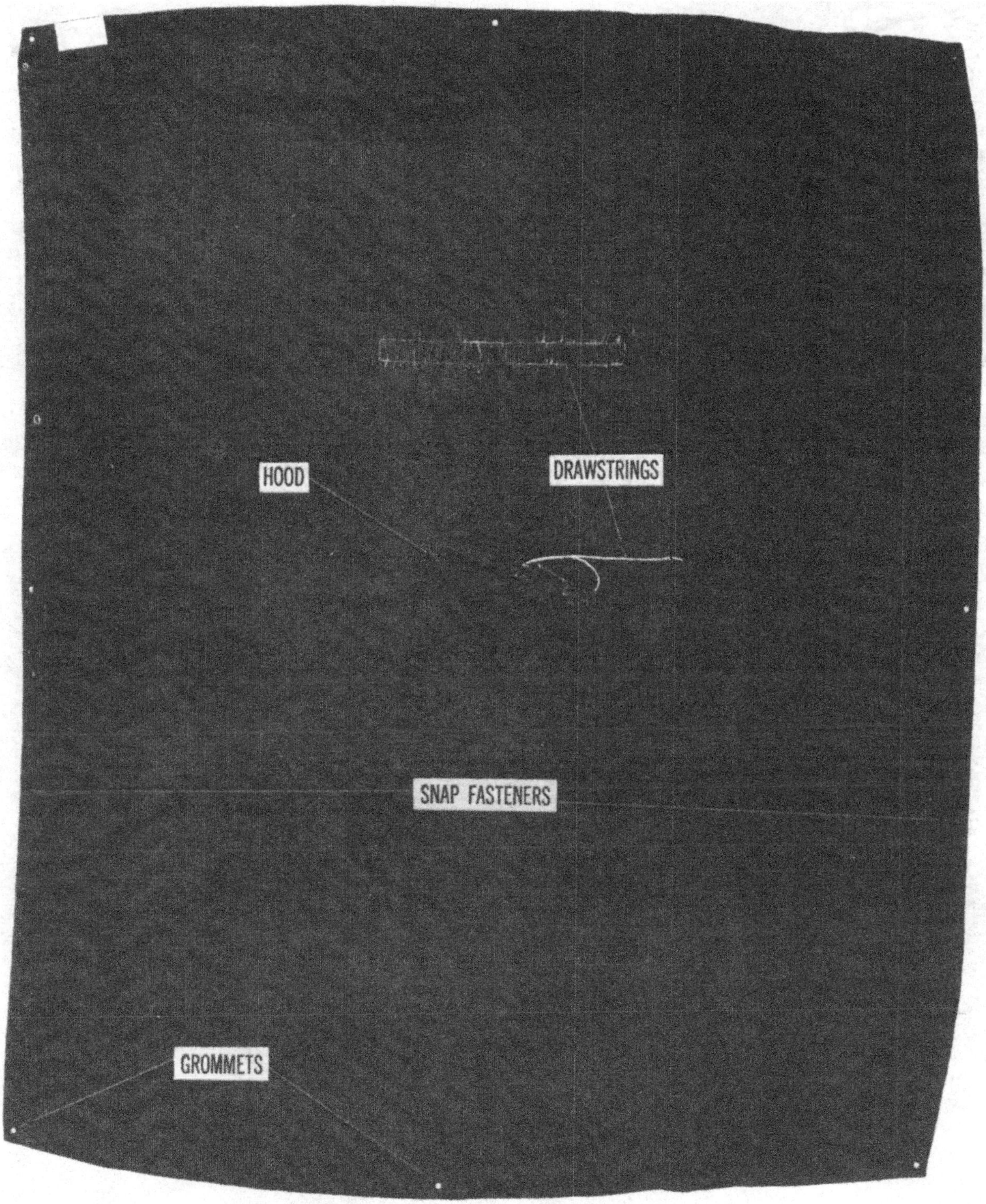

Figure 5. Lightweight poncho.

Figure 6. Poncho used as a rain garment.

the long sides parallel and the short sides slightly curved. The hood and the neck opening are located at the center of the rectangle. The poncho is also available in camouflage pattern (app. B).

b. Use. Use the poncho as a rain garment (fig. 6) by wearing it as a raincape with the arms inside or wearing it with the arms outside for freedom of movement. If the poncho hood is to be worn, adjust the hood drawstring to fit. Wear the helmet and/or liner over the hood. Adjust waist drawstring by fitting poncho around waist and tying drawstring at back. Fasten snap fasteners together on each side of the poncho. To prevent poncho from flapping in high winds, make certain all fasteners are fastened. Other uses for the poncho are given in paragraphs 23 and 37.

c. Fold and Carry. The poncho can be folded and carried with the lightweight load-carrying equipment (para 28) or carried over the individual equipment belt. Carry it in a duffelbag or inside the field pack if space permits. Fold the poncho in any appropriate manner, but take care not to damage it in folding or by pressure against sharp or rough objects in the bag or pack. Figure 7 shows how to fold the poncho for carrying with the load-carrying equipment. Figure 8 shows how to fold the poncho when it is to be carried over the individual equipment belt.

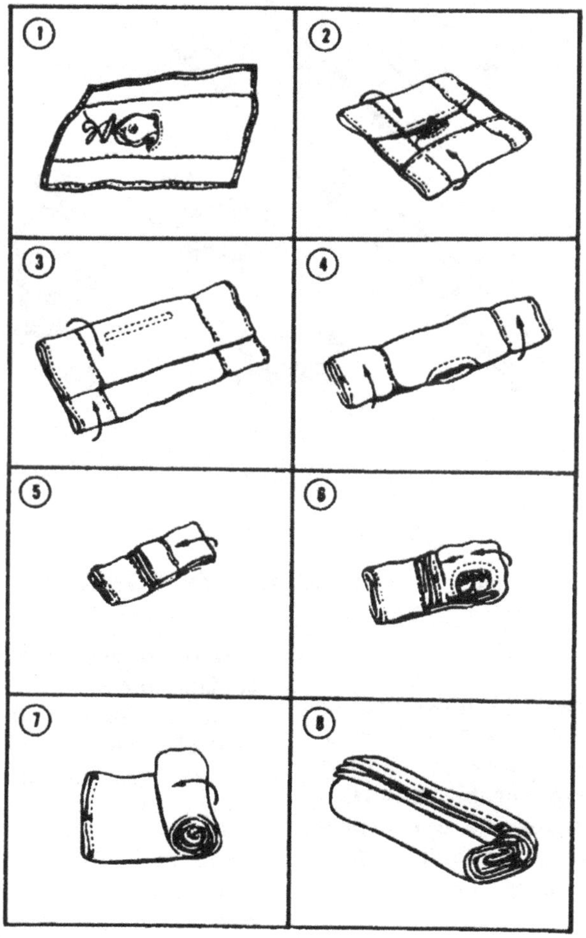

Figure 7. Steps in folding poncho for carrying with load-carrying equipment.

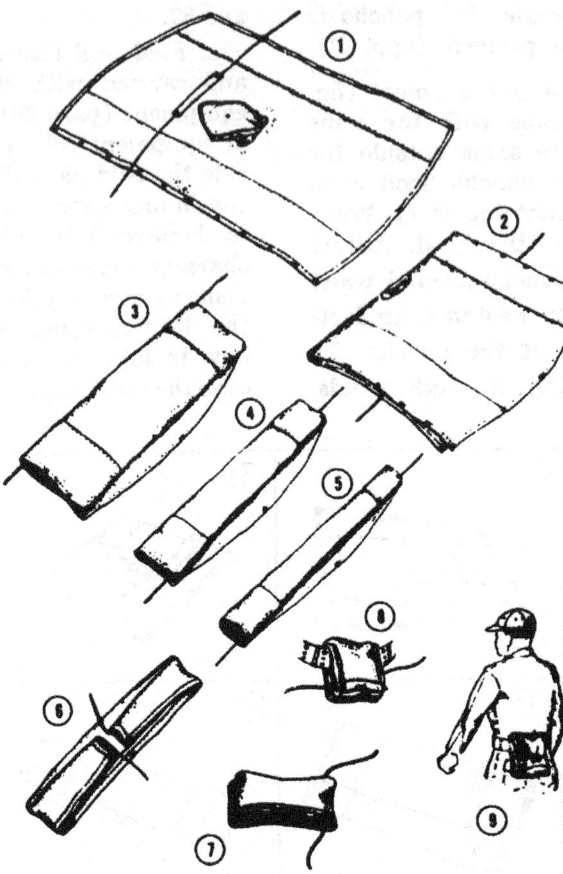

Figure 8. Steps in folding poncho for carrying over individual equipment belt only.

Section II. FOOTWEAR COMPONENTS

15. General

Hot weather footwear is made of quick-drying materials. The sock material maintains resiliency even when wet. The boots are lightweight and are constructed of highly vapor-permeable materials to allow evaporative cooling for the upper portions. Eyelets are used when practicable to permit water drainage and to help ventilate the feet.

16. Socks

The olive-green cushion sole socks are the standard wool stretch-type socks.

17. Hot Weather Tropical Boots

a. Description. The spike-resistant hot weather tropical boots (fig. 9) have direct molded soles, nylon duck quarters, two screened vents located at each inside arch, and removable ventilating plastic insoles. The outsole has two systems of cleats projecting at different levels. The heel also has a similar cleat. A stainless steel protective plate is incorporated within the outsole.

b. Use. Wear the boots over the cushion sole socks in hot damp environment. The vents permit water to drain from the boots without your having to take them off. The vent screens prevent insects from getting inside. The ventilating plastic insoles provide an air circulatory space underneath the feet. The built-in stainless steel plates aid in preventing nails, spikes, or other sharp objects (punji) from penetrating the bottom of the boots. The boot cleats are designed to increase traction and reduce the tendency for mud to cling to the outsoles.

Figure 9. Hot weather tropical boot (panama sole).

Section III. FITTING OF HOT WEATHER CLOTHING

18. Body Clothing

a. Sizes. The sizes of most body clothing items are stated in terms of length and width. The lengths are specified as *short, regular,* and *long*; the widths are specified as *extra small, small, medium, large,* and *extra large.* Sizes available in hot weather clothing are shown in tables 1 and 2.

Measurements. To determine the proper size clothing, it may be necessary to obtain height and chest measurements. To measure—

(1) *Height.* Measure from the floor to the top of the head, with the subject standing erect.

(2) *Chest.* Place a tape measure well up under the arms and over the shoulder blades on a horizontal line, using the floor as a level. Take snug measurements from the center of the back with the subject standing naturally and *not* with the chest fully extended.

c. Fitting. Hot weather clothing is designed to be loose fitting and should be selected to fit comfortably. Tight fitting clothing reduces airflow for ventilation and cooling. Proper sizes to select for try-on are given in (1) and (2) below. If an item is not issued in different *lengths,* choose the size on the basis of *width* alone.

(1) *Chest measurements.*
If chest measures—

Inches	*Select— (Size)*
32 and below	x-Small
33 through 36	Small
37 through 40	Medium
41 through 44	Large
45 and above	x-Large

(2) *Height measurements.*
If height measures—

Inches	*Select— (Length)*
66 and below	Short
67 through 70	Regular
71 and above	Long

(3) *Waist measurements.* If waist measures—

Inches	*Select— (Size)*
26 and under	Extra Small
27 through 30	Small
31 through 34	Medium
35 through 38	Large
39 and over	Extra Large

(4) *Inseam measurements.* If inseam measures—

Inches	*Select— (Length)*
29 and under	Short
29½ through 32	Regular
32½ and over	Long

19. Headgear

a. Sizes. Sizes for hot weather headgear correspond with established cap sizes. Sizes available are shown in table 2.

b. Fitting. Check the size of the subject's standard uniform cap and select the same size field cap or tropical hat with detachable head net. Try on various sizes until a comfortable fit is obtained.

20. Footwear

a. Sizes. Cushion sole socks are issued in *small, medium,* and *large.* Sizes for tropical combat boots usually correspond with the established sizes for standard footwear (TM 10-228).

b. Fitting. Select the cushion sole socks according to the individual's standard sock size. Check the size of the subject's standard boot and select the same size tropical combat boot for initial try-on. Check the fit of the boots while the subject is standing with weight evenly distributed on both feet. Arch supports or other orthopedic appliances required must be worn while trying on and fitting boots. Boots selected should be comfortable without a "breaking in" period. Details for the fitting of footwear are contained in TM 10-228.

Table 1. Body Clothing Sizes

ITEM	Size								
	Length	Width							
		xx-Small	x-Small	Small	Medium	Large	x-Large	xx-Large	xxx-Large
Undershirt, cotton, short sleeve		x	x	x	x	x	x	x	x
Drawers, cotton (boxer)			x	x	x	x	x	x	x
Coat, hot weather, man's	Short		x	x	x	x			
	Regular		x	x	x	x	x		
	Long			x	x	x			
Trousers, men's combat, tropical	Short		x	x	x	x			
	Regular		x	x	x	x	x		
	Long			x	x	x			
Shirt, sleeping man's, nylon/triacetate			x	x	x	x	x	x	

Table 2. Footwear, Headgear, and Miscellaneous Clothing Sizes

Item	Size
Boots, hot weather tropical (spike-resistant).	Same as for leather combat boots (51 sizes).
Socks, olive green cushion sole.	Small, medium, and large.
Cap, hot weather	6⅜, 6½, 6⅝, 6¾, 6⅞, 7, 7⅛, 7¼, 7⅜, 7½, 7⅝, 7¾, 7⅞, (13 sizes).
Hat, tropical, with detachable head net.	Same as cap, hot weather above.
Neckerchief	One size (36" x 24").
Poncho, lightweight	One size.

CHAPTER 3

HOT WEATHER SLEEPING EQUIPMENT

21. General

A means to shield the body from the wet ground is essential in hot tropical areas, especially during the rainy season. Using lightweight hot weather sleeping equipment with accessories appropriate to the environment, the soldier can rest comfortably under various hot weather conditions.

22. Jungle Hammock

a. Description. The lightweight jungle hammock (figs. 10 and 11), made of a single layer of nylon fabric, is 8 feet 4 inches long and 2 feet 8 inches wide. One 14-foot long, ½-inch diameter braided cord is passed through the drawcord tunnel at each end. A rubber drip

Figure 10. Jungle hammock.

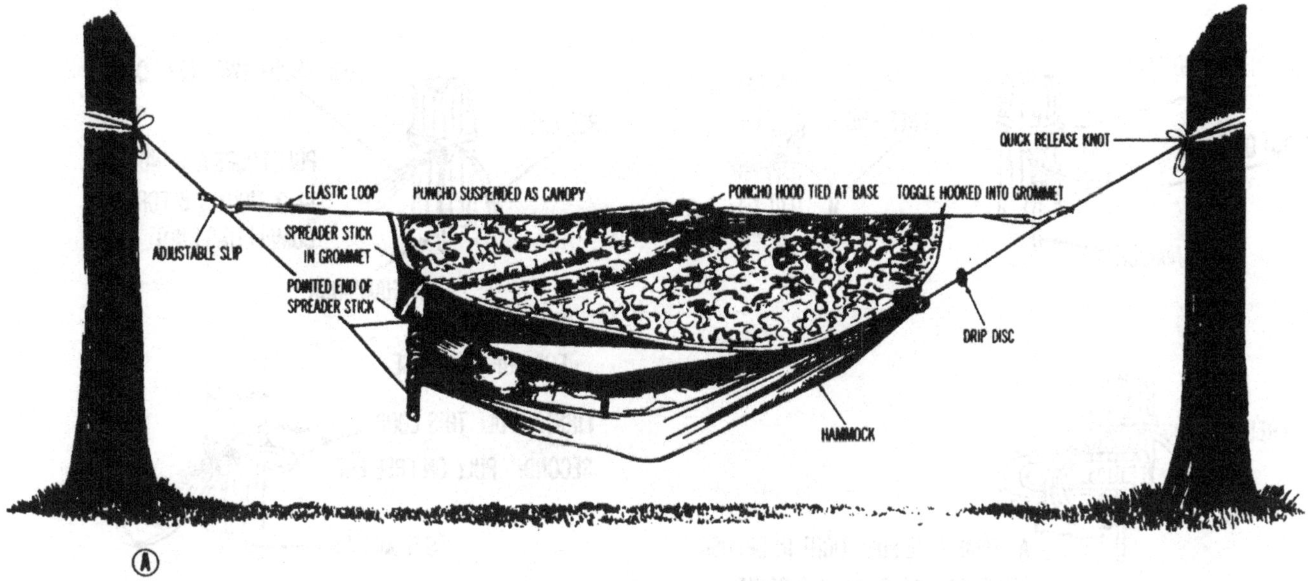

Figure 11. Using end spreader stick.

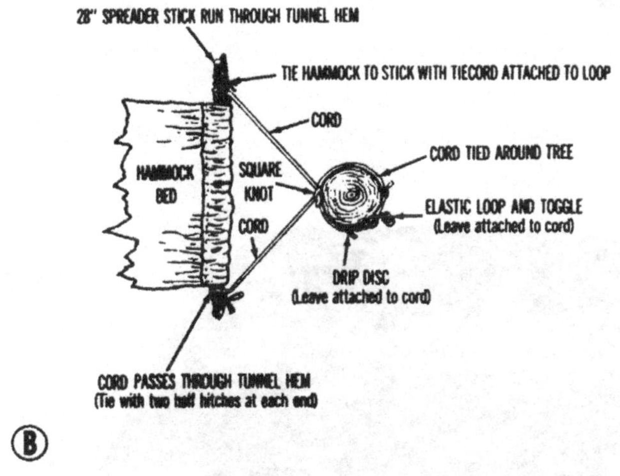

Figure 11.—Continued.

disc is provided at each end of the cord. A nylon tab and tiecord are located at each corner of the hammock to accommodate spreader sticks, if desired (B, fig. 11).

b. Use. Suspend the hammock between two trees at least 10 feet apart using a quick-release knot (C, fig. 11) to secure the cords around the tree trunks. If a rain canopy is desired, suspend the lightweight poncho (para 14) over the hammock by using the poncho adjustable slip and elastic loop suspension assemblies (A, fig. 11) at each end. To do this—

(1) Put the toggles through the center grommets in each end of the poncho.

(2) Adjust the slips so that the ridge line of the poncho is nearly a straight line when the hammock is suspended.

(3) Cut two spreader sticks not over 5 feet long and sharpen the ends.

(4) Put a stick between the corner grommets at each end of the poncho with the sharpened ends through the grommets, making sure the sticks are *under* the hammock suspension cords.

(5) Position the rubber drip discs just above the hammock suspension rope knots at each end of the hammock.

(6) Tie off the poncho hood to prevent entry of rain through the hood opening.

c. Carry. Fold the hammock in accordion-like folds with one hammock suspension cord tucked within the folds and the other wrapped around the folded hammock.

23. Poncho Liner With Poncho

a. Description. The poncho liner (fig. 12) conforms in size and shape to the lightweight poncho (para 14). It is camouflage patterned and is made of polyester batting sandwiched between two panels of light nylon fabric. The liner is equipped with eight tie tapes.

b. Use. Use the liner as a blanket or as a sleeping bag when used in conjunction with the

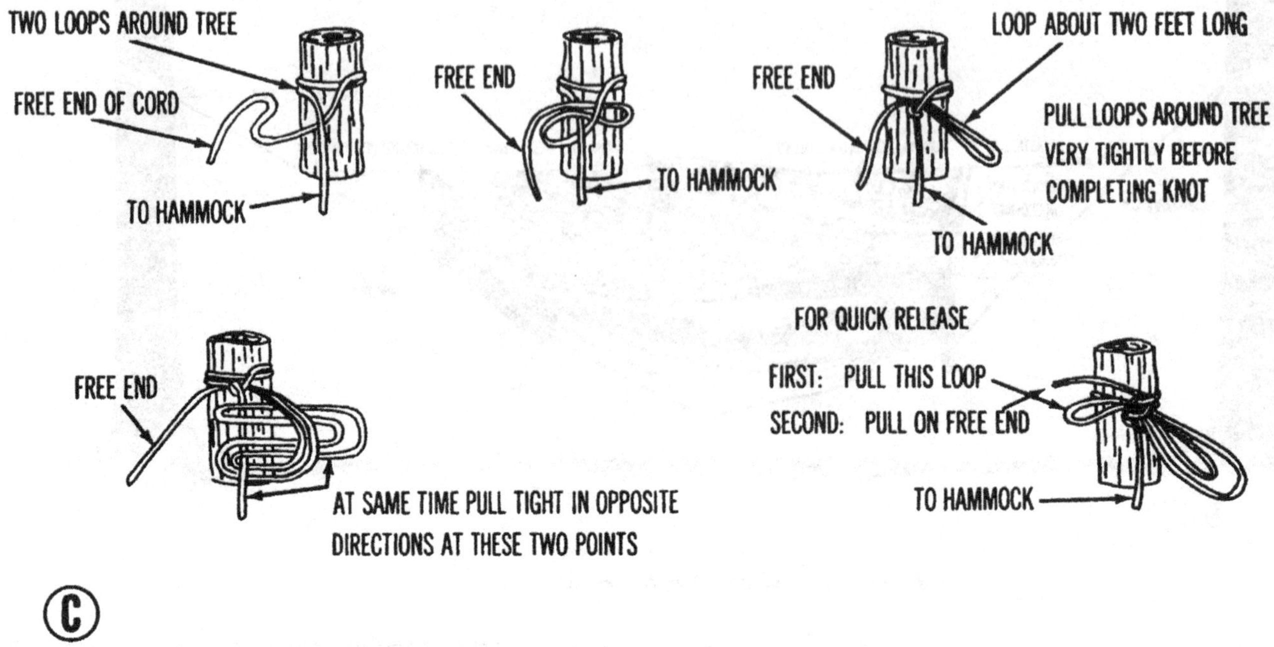

Figure 11.—**Continued.**

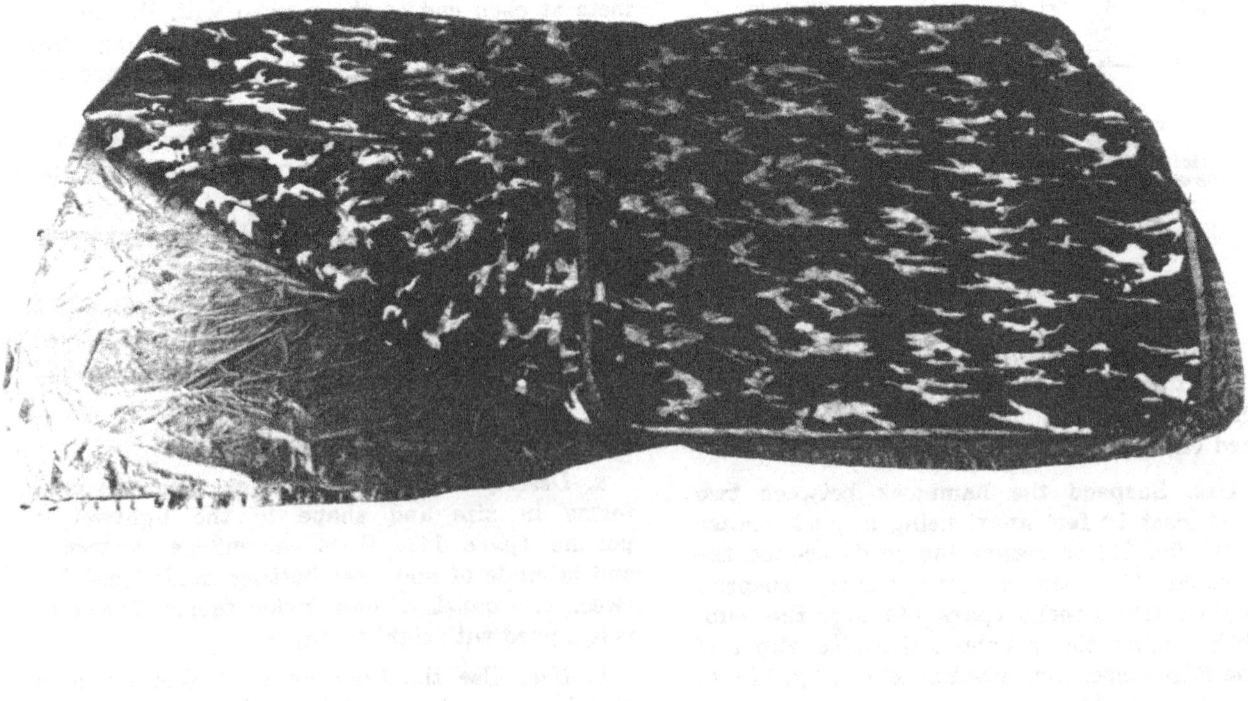

Figure 12. Poncho liner with poncho.

poncho (fig. 13). To attach the poncho liner to the poncho (fig. 14) proceed as follows:

(1) Spread the poncho on the ground, making sure that the hood opening is tightly closed and on the underside (A).

(2) Place the liner on the poncho, matching the tie tapes on the liner with the grommets on the poncho, and tie the poncho and liner together (B).

(3) Fold the poncho and liner in half lengthwise and fasten all snap fasteners along the long side (C). *Do not fasten the snap fasteners when the poncho sleeping bag is used in combat areas, because it cannot be opened quickly.*

(4) Tuck the foot end under to keep feet from protruding (D).

24. Multipurpose Net as a Hammock

The multipurpose net (para 31) may be used as a hammock (fig. 15). To do this, lay the net out full length on the ground, using the selvaged side of the net as the sides of the hammock. Gather approximately 9 inches of the ends of each short side of the net and wrap and tie each end with a 14-foot cord. Using the other end of the 14-foot cord, tie each end of the net to a tree using a quick-release knot (C, fig. 11), making sure that the ties are at least 3 feet high to compensate for sag. Pull the cords tight. Use one 12-inch length of wood as a spreader bar by tying the bar approximately 12 inches from the gathered portion of the head end of the net with nylon cord. Do *not* thread the ties through the mesh of the net. Figure 16 illustrates a three-

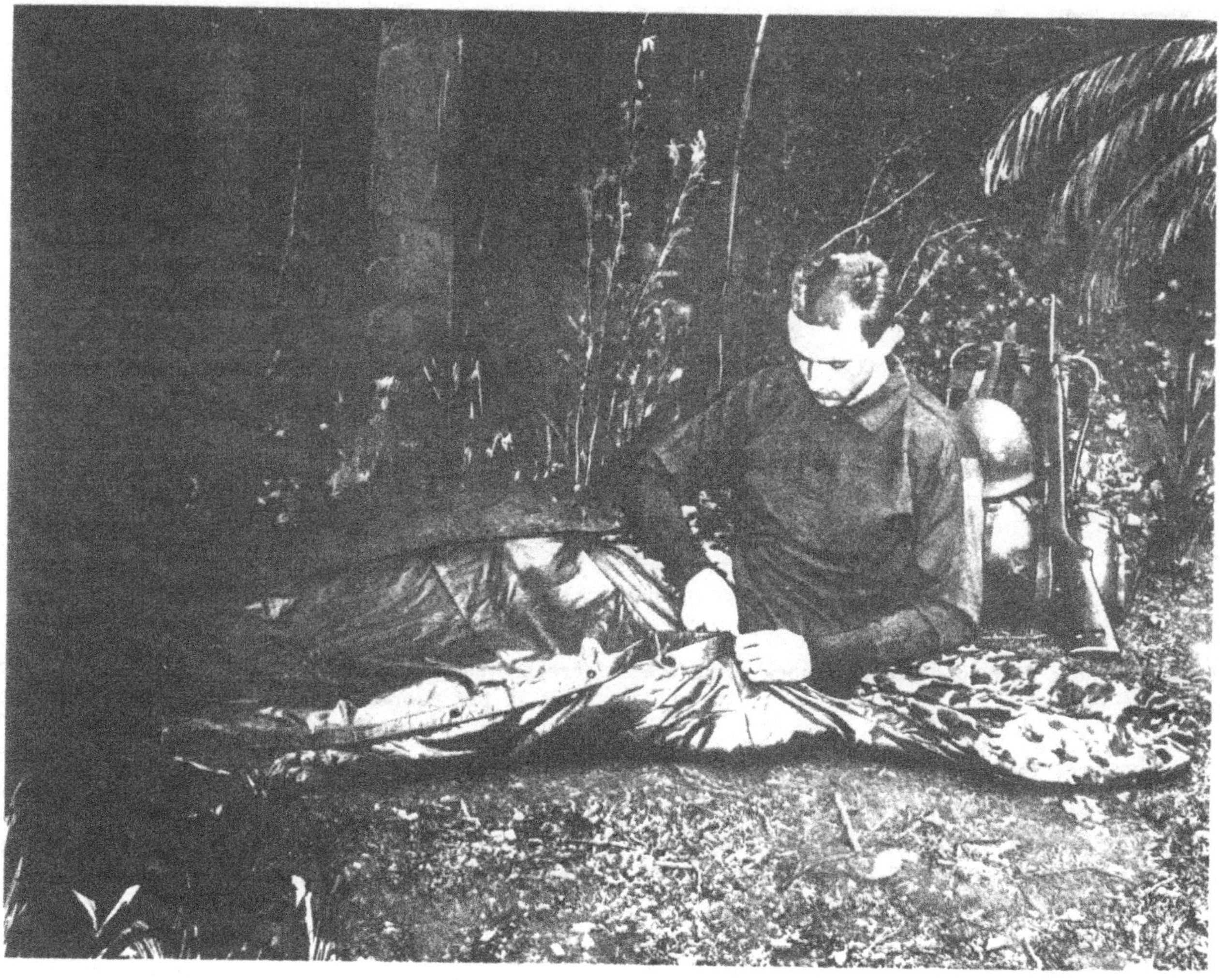

Figure 13. Poncho sleeping bag.

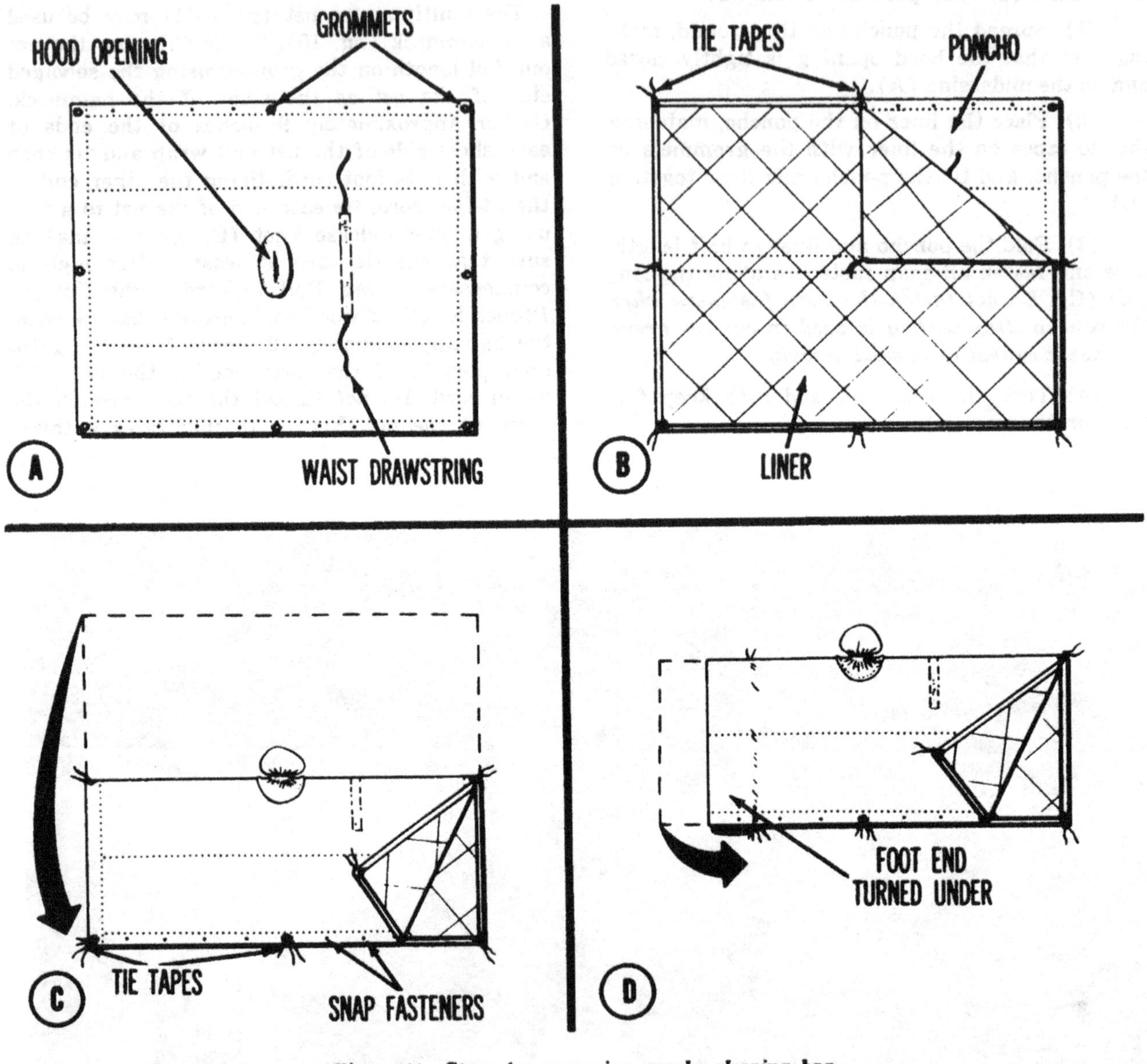

Figure 14. Steps for preparing poncho sleeping bag.

cornered and a four-cornered hammock using the multipurpose net.

25. Sleeping Accessories

a. Pneumatic Mattress.

(1) *Description.* The pneumatic mattress (fig. 17) is made of a coated fabric and is shaped to conform generally to the sleeping bag.

(2) *Use.* Use the mattress under sleeping equipment to keep it dry and to add warmth and comfort. Do not overinflate; overinflation decreases sleeping comfort. Sit on the mattress to test it for proper inflation and comfort. If the mattress is properly inflated, the buttocks will barely touch the ground. Before inflating the mattress, make sure there are no protruding thorns, rocks, or sharp objects which may puncture it. The mattress can also be used as a raft or floating device to transport equipment over water (fig. 18) or as an aid to swimmers.

b. Sleeping shirt.

(1) *Description.* The olive green nylon/tria-

Figure 15. Using multipurpose net as a hammock.

cetate sleeping shirt (fig. 19) is shrink resistant and resembles a pullover sweater. It has a convertible collar with a two-button placket closure and full length sleeves with knitted cuffs.

(2) *Use.* Wear the sleeping shirt under the tropical combat coat for added warmth on cool nights or in high elevations where the air tends to cool off markedly after sundown. When the tropical combat coat is wet and the air is cool, wear the sleeping shirt in place of the coat to avoid chill.

Figure 16. Using multipurpose net as three- and four-cornered hammocks.

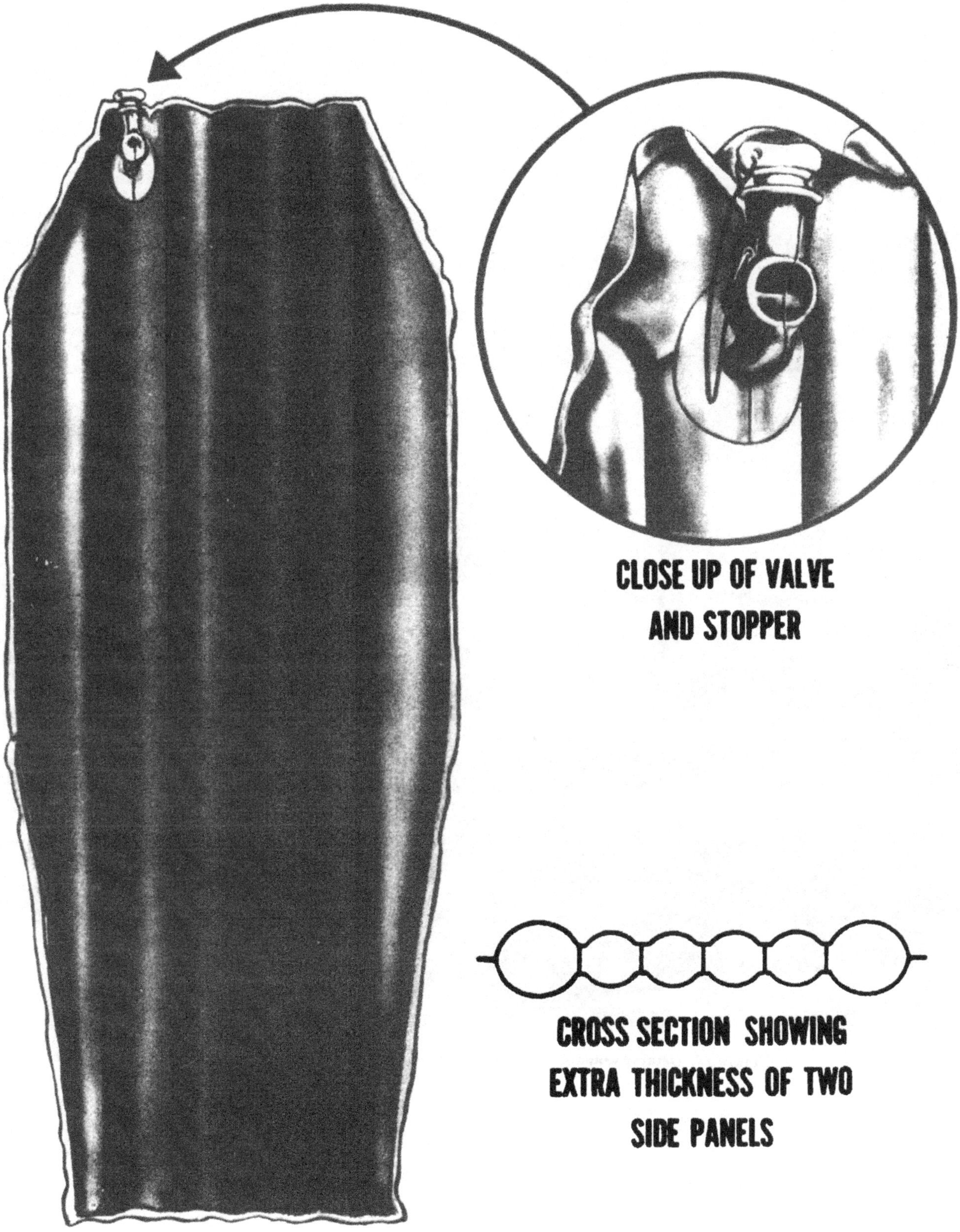

Figure 17. Pneumatic mattress.

Figure 18. Using pneumatic mattress to transport equipment over water.

Figure 19. Sleeping shirt.

CHAPTER 4

HOT WEATHER LOAD-CARRYING EQUIPMENT

26. Tropical Rucksack With Liners

a. Description. The tropical rucksack frame is made of spring steel. The fabrics and webbing are made of water-resistant, fast-drying nylon. The rucksack (fig. 20) consists of a pack and three outside pockets. Each pocket has an adjustable closure and snap fasteners which allow quick access. Equipment hangers are located above each pocket; each hanger can accommodate equipment with double end hooks of sliding keepers. The pack top flap (which serves as a waterproof cover for the pack) is also a pocket. This top flat pocket has a nylon hook-and-pile closure. The two loops on the bottom and two on each side of the pack are used to attach items to the pack. An equipment hanger, with tiedown strap, is also located on each side of the pack. Waterproof liners (bags) are provided for the pack and each outside pocket. The tropical rucksack is more desirable than the standard lightweight rucksack because it weighs less and has fewer projections to snag on the underbrush.

b. Use.

(1) *Rucksack.* Use the rucksack to carry individual clothing, equipment, and rations. The backstrap, which flexes the metal strap at the bottom of the frame, should be kept tight with a bow of approximately 2 inches. This bow will keep the pack away from the wearer's back for more comfort and will allow more air circulation to the back. Because the backstrap tends to loosen with wear, it should be adjusted at frequent intervals. The side frame straps should be kept tight in order to form a vertical curve at the wearer's back and to keep the frame from collapsing outward away from the back. The pack also can be used without the frame.

(2) *Liners.* Waterproof liners for the tropical rucksack are provided in two sizes—a small size for use in each of the three outside pockets; and a large size for use in the pack. Each liner has an attached tie cord. Use the liners to protect the contents of the rucksack pack and pockets from moisture. The liners also may be used as flotation devices for either the rucksack or the individual.

Figure 20. Tropical rucksack.

(a) *To keep equipment dry.* Insert the small liners into each of the three outside pockets, insert the large liner into the pack, and pack the liners. Close the liners by wrapping the tie cords around the top of the liners and tying the tie cords with a double bow or a square knot.

(b) *To give support in water.* Flatten the mouth of the liner and fold the flattened portion in half approximately four times. Insert a finger within the folds to make an opening for the purpose of inflation. Inflate the liner approximately

50 percent. Turn the folds 3 inches, making sure the tie cord remains on the outside. Wrap the tie cord tightly around the folded portion and tie a double bow knot or a square knot. Place each inflated bag in the proper pocket and pack, and close and tie down the flaps.

27. Lightweight Rucksack With Riveted Frame

Details on the standard lightweight rucksack with riveted frame (fig. 21) are covered in TC 10-8.

28. Lightweight (Nylon) Load-Carrying Equipment

Components of the lightweight load-carrying equipment (figs. 22 and 23) are similar in design to components of the standard M-1956 individual load-carrying equipment. Because nylon materials and aluminum hardware are lightweight, they are used where possible in lieu of cotton materials and brass or steel hardware. The nylon materials are water resistant, retain less mois-

Figure 21. Lightweight rucksack with riveted frame.

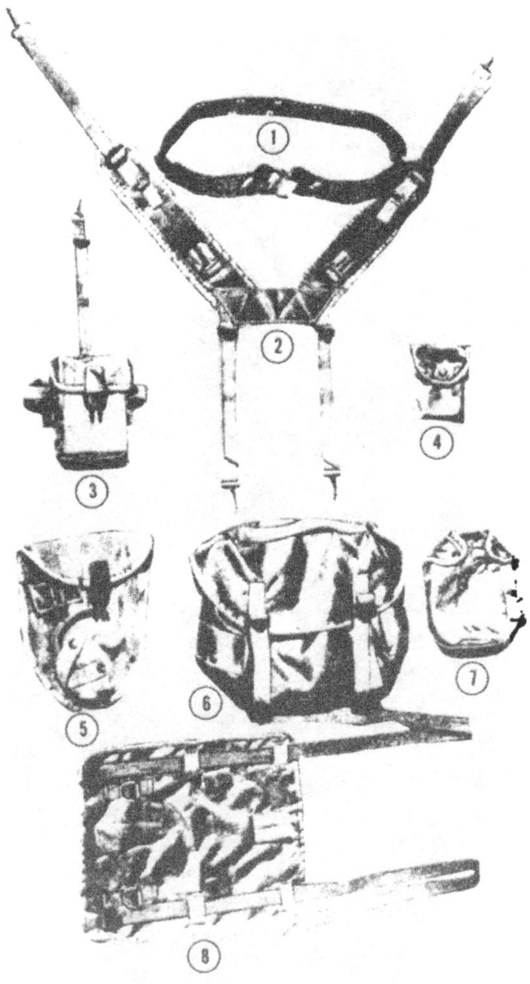

1 Individual equipment belt
2 Suspenders
3 Small arms ammunition case
4 First aid packet or compass case
5 Intrenching tool carrier
6 Combat field pack
7 Canteen cover
8 Sleeping equipment carrier

Figure 22. Components of lightweight load-carrying equipment.

ure, and dry faster. Details on description, fitting, and principle of use of components of the standard M-1956 individual load-carrying equipment, which are the same as those for the lightweight load-carrying equipment, except for differences noted, are contained in FM 21-15.

29. Multipurpose Net Used for Load Carrying

The multipurpose net (para 31) may be used for carrying loads (fig. 24). To carry the load without using the hands (A), lay the net out full

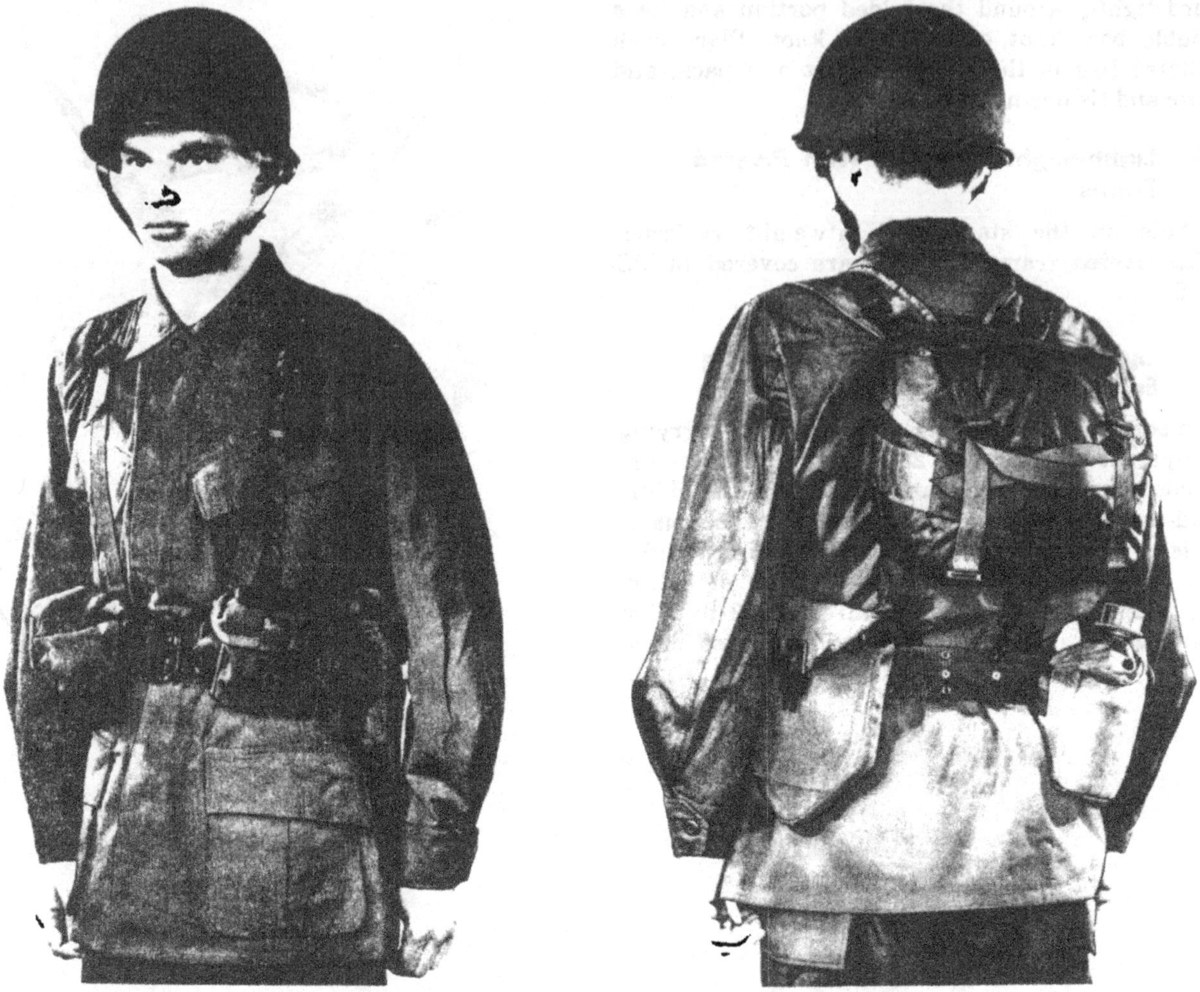

Figure 23. Wearing lightweight (nylon) load-carrying equipment.

length on the ground and place the load to be carried in the middle of the net. Pull each side of the net over the load and cross over the loose ends on each side of the load. Place the load on the back. Draw the loose end, which is against the back, up and around the load and over the shoulder. Draw the other loose end under the arm to form a carrying strap. Pull and tie the ends up tight so that the load will ride high on the shoulder. Repeat the procedure for the opposite side. To carry a load using the hands, lay the load to be carried in the center of the net, gather the ends together, and carry the load as illustrated in B, figure 24.

30. Grenade Carrier Vest

a. Description. The M-79 grenade carrier vest (fig. 25) is issued in small and large sizes. The front is made of nylon duck and the shoulders and back are of nylon mesh to provide increased ventilation to the wearer. The front closure is a nylon hook-and-pile tape fastener reinforced with two snap fasteners. A nylon strap and quick-operating buckle provide fitting adjustment and closure at the back. There are 24 grenade pockets on the vest, 12 on each front side. Each grenade is held in place in the pocket by a snap fastener.

b. Use. The vest is designed to be worn over the armor vest and suspenders of the load-carrying equipment. It enables the wearer to carry 24 grenades comfortably and securely. Each grenade can be removed with one hand.

Figure 24. Two methods of carrying loads using multipurpose net.

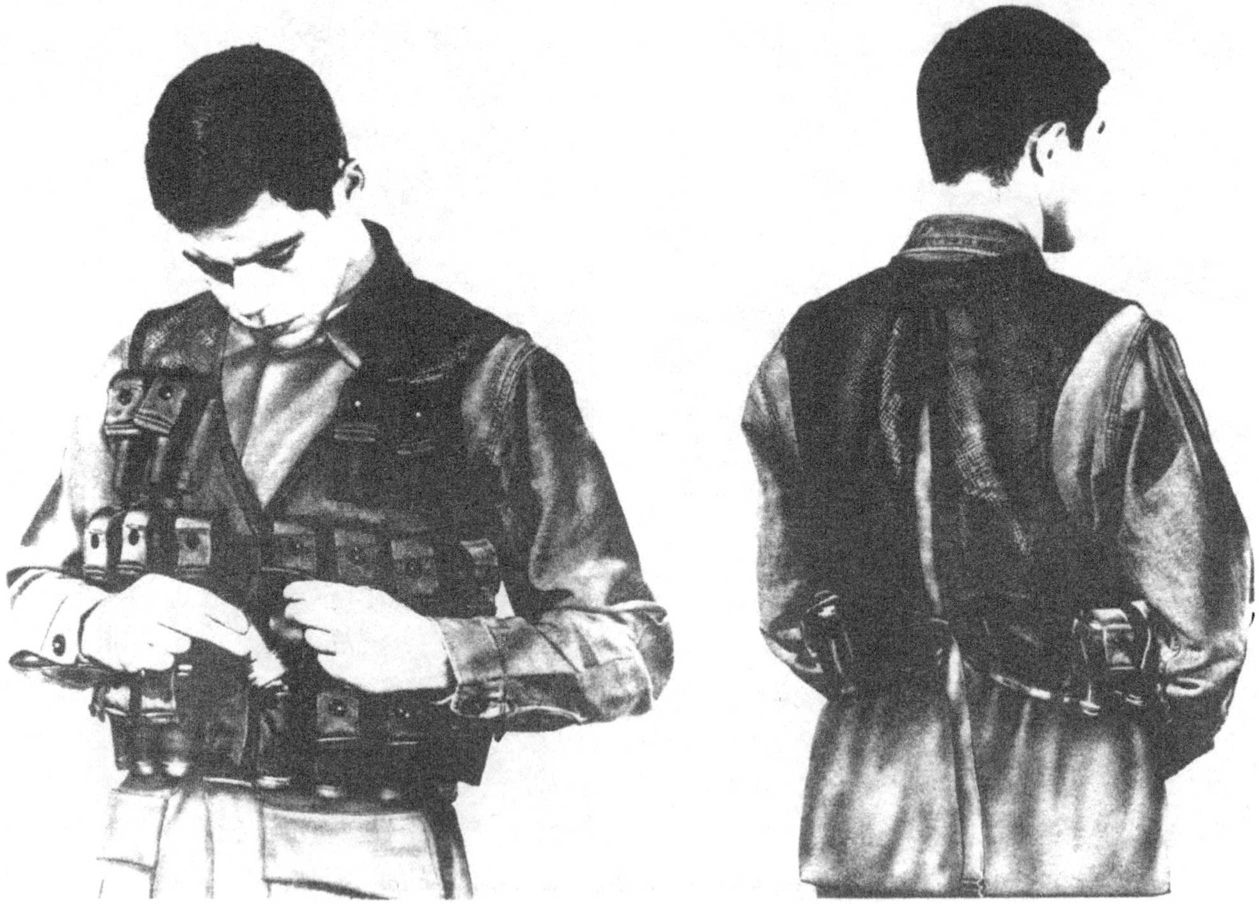

Figure 25. M-79 grenade carrier vest.

CHAPTER 5

MISCELLANEOUS HOT WEATHER EQUIPMENT

31. Multipurpose Net

a. Description. The multipurpose net (fig. 26), constructed of lightweight, olive-green nylon, weighs 1 pound, and is approximately 5 feet wide by 9 feet long. The long sides have nylon selvage and the short sides have melted ends. Included with each net are two cords, approximately 14 feet long, with fused ends. When folded, the net can be carried in the rucksack or in the trousers pocket, or it can be attached to the individual belt in a similar manner described for the poncho (para 14) or to the rucksack harness (para 26).

b. Use. In addition to its use as a hammock (para 24) and as a means of carrying loads (para 29), the multipurpose net has many other uses as shown in figure 27.

32. Canteens

a. Plastic Water Canteen.

(1) *Description.* The plastic (polyethylene) water canteen (fig. 28) holds 1 quart. The canteen, which fits into the standard canteen cup, is olive drab, has a plastic screwcap with an attaching strap, and is carried in the standard canteen carrier.

(2) *Use.* Use the canteen for water only. When required, replace the cap with the M-1 drinking device; this allows water to be consumed from the canteen while the user is wearing a protective mask (TM 10-277).

CAUTION

Do *not* apply the canteen to an open flame or to burner plates.

b. Collapsible Water Canteen.

(1) *Description.* The collapsible water canteen (fig. 29) consists of a plastic (ethylene-vinyl acetate copolymer) bladder with the neck at one corner. It measures 7 by 7 by 3 inches and holds a minimum of 2 quarts. The cover is made of water-repellent nylon and has an acrylic pile lining. A pocket for holding water purification tablets is sewed to the front of the cover. The canteen is puncture- and tear-resistant and can be easily inserted or removed from the cover when empty or full.

(2) *Use.* Carry the canteen cover with the shoulder strap or attach it to the individual equipment belt. The canteen can be filled from a running or still source of water (fig. 30). To fill the canteen, remove the bladder and unscrew the cap. Blow into the neck of the bladder to extend it fully and insert the correct number of water purification tablets (as directed on the bottle of tablets). Float the empty bladder on the side and hold the neck of the bladder with approximately one-third to one-half of the neck opening under water. Do not totally submerge the neck opening. Allow the bladder to slowly submerge as it fills. If the bladder collapses, remove it from the water and allow air to enter it; then proceed as described above. To fill the canteen quickly, use a scooping motion with the neck opening partially under water as described above. As water is consumed, or when the canteen is partially filled, squeeze the bladder to force water into the neck; then replace the cap tightly. This procedure will reduce the sloshing noise. To help keep the water in the canteen cool, kept the canteen cover lining wet. When required, replace the cap with the M-1 drinking device; this allows water to be consumed from the canteen while the user is wearing a protective mask (TM 10-277).

33. Flotation Bladder Assembly

a. Description. The flotation bladder assembly (fig. 31) consists of a 5-quart capacity vinyl film bladder (A) and a nylon cover or carrier (B).

(1) *Bladder.* The bladder measures approximately 13½ by 10½ inches and has a canteen-like neck with a cap equipped with a rubber gasket. A removable strainer filter is inserted in the neck of the bladder.

(2) *Cover or carrier.* The cover or carrier

Figure 26. Multipurpose net.

Figure 27. Various uses for the multipurpose net.

Figure 27.—Continued.

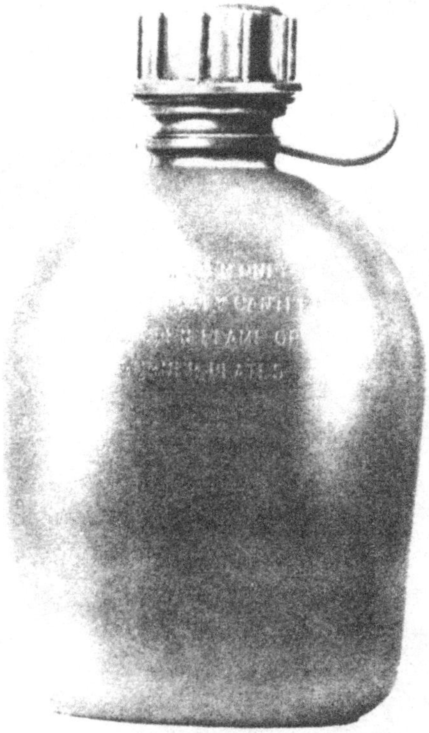

Figure 28. Plastic waer canteen.

measures approximately 13½ by 10½ inches and is equipped with retainer loops and tiedown cords at each corner. An opening with snap fastener closure is provided at the long side of the carrier and a fabric funnel-shaped pouch is sewn onto the neck opening. A pocket for holding water purification tablets is sewed to an upper flat side of the carrier. Directions for inflating the bladder are located on one flat side of the carrier; directions for filling the bladder with water are located on the opposite side.

b. Assembly. To insert and attach the bladder to the carrier, refer to figure 31 and proceed as follows:

(1) Open side opening snap fasteners and extend side opening flap outward.

(2) Fold bladder twice lengthwise in 3½-inch increments. Then roll the bladder widthwise up to the neck to form a 1½-inch roll.

(3) Insert the roll, neck first, through the side opening into the carrier.

(4) Continue by passing the bladder neck with or without cap through the carrier mouth opening.

NOTE

If passing the cap with it attached to the bladder neck is difficult, remove the cap and pass it through the carrier mouth opening first; then pass the bladder neck through.

(5) With the bladder neck fully drawn through the carrier mouth opening, draw the mouth pull-up cord tightly around the bladder neck and tie a square knot. This operation locks the bladder neck in the carrier mouth.

(6) Unroll and unfold the bladder inside the carrier, making sure it is flat and free of wrinkles.

(7) Reinsert side opening flap and close snap fasteners. Put cap on, if necessary.

(8) Fold funnel-shaped pouch over cap and attach long closure strap to short closure strap.

c. Use. Use the flotation bladder assembly as a flotation device for the individual or his equipment or as an expedient canteen.

(1) *Flotation device.* Inflat the bladder within the carrier by holding it vertically and blowing into the bladder neck. Attach the carrier to personnel or equipment by using the retainer loops and tiedown cords located at each corner of the carrier.

(2) *Canteen.* Fill the bladder from a still source of water in the same way that the 2-quart collapsible canteen is filled (para 33*b*). To fill the bladder from a stream—

(*a*) Remove the cap and filter screen, insert correct number of water purification tablets (as directed on the tablet bottle), and replace the filter screen.

(*b*) Squeeze air from the bladder.

(*c*) While gripping neck with middle fingers and thumbs, hold the funnel-shaped pouch open with the forefingers and thumbs, and repeatedly dip the bladder neck vertically under water. This operation keeps the neck covered with water inside the pouch.

d. Folding. When not in use for flotation purposes or for carrying water, the bladder assembly can be folded for storage or carry. To fold, refer to figure 32 proceed as follows:

(1) Lay the assembly (carrier with bladder inserted) flat on a suitable surface with the small pocket side on the bottom (A).

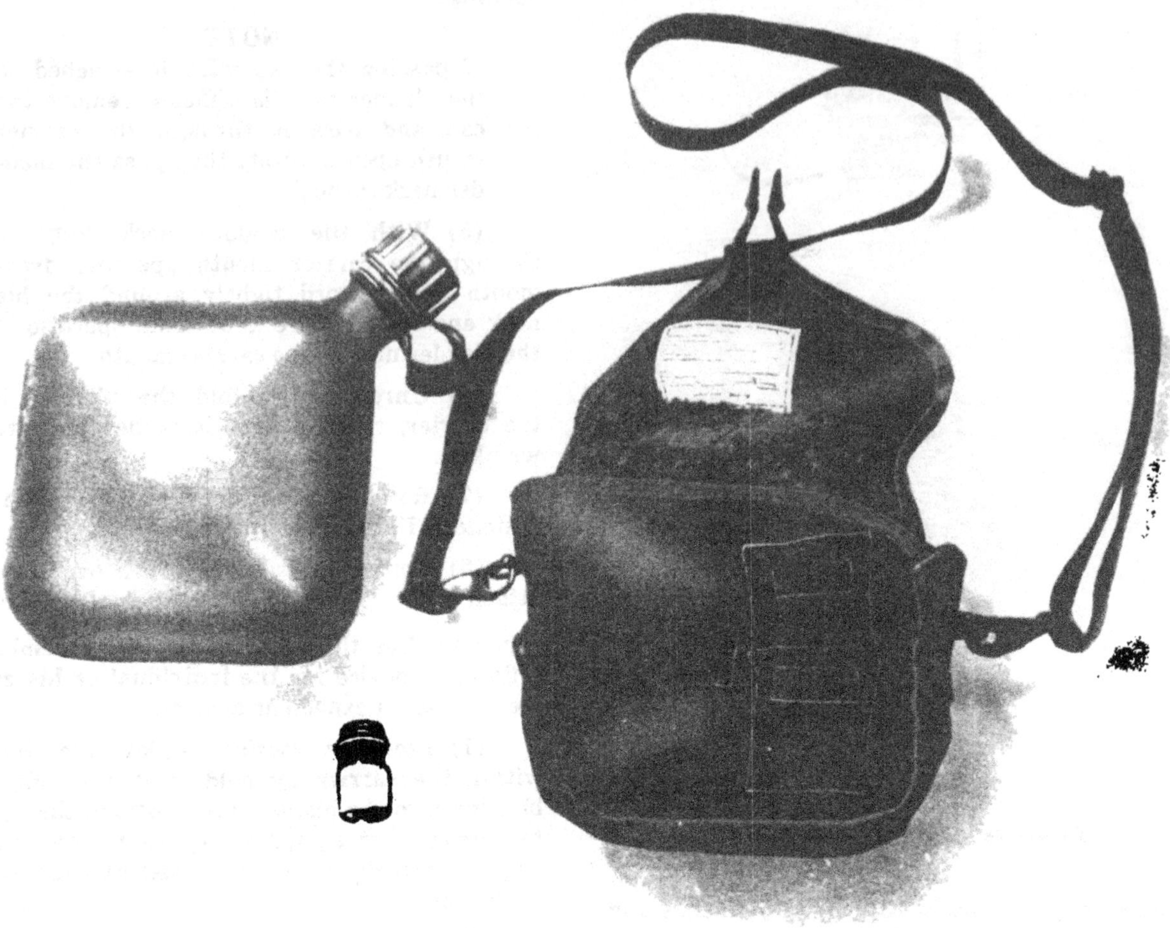

Figure 29. Collapsible 2-quart water canteen.

(2) With the funnel-shaped pouch and long closure strap fully extended, place the tiedown cords and retainer loops toward the center of the carrier (B).

(3) Fold each side of the assembly inward *width wise* so that the folds overlap the center of the assembly by approximately 1 inch (C and D).

(4) Next, fold the assembly twice *lengthwise* starting at the bottom of the folded assembly (E and F).

(5) Pull the funnel-shaped pouch over the complete folded assembly (G). This operation results in the folded assembly being completely inclosed in the pouch with the bladder neck and cap exposed.

(6) Insert short closure strap inside the pouch, if necessary.

(7) Twist long closure strap 180° so that the pile tape on the long closure strap will engage the hook tape located underneath the carrying loop (H).

34. Machete With Plastic Sheath

a. Description. The M-1942 machete (fig. 33) is a straight-backed, 18-inch knife with a smooth plastic handle. It is an effective tool for cutting through underbrush and an effective weapon when silence is imperative.

b. Carry. The machete is best carried in its sheath, which should be attached to the individual equipment belt on the left side of the body. When carried in this manner, the edge is forward and the machete can be drawn easily without danger of cutting the legs or body. *The machete should be sheathed when not in use.*

Figure 30. Filling collapsible water canteen from a stream.

c. Use. The machete is a cutting instrument that depends on velocity and angle rather than weight for its effectiveness. Never grasp the machete handle tightly as if it were a hatchet, and never hold it so loosely that it is out of your control. Obtain maximum velocity by gripping the handle firmly with the thumb and first two fingers and loosely with the last two fingers. When a strike is made, the fingers and wrist should snap the machete forward in a whip-like motion. Just before the blade strikes its target, the last two fingers should be tightened on the handle to give added strength and velocity to the blade. This technique will keep the machete from slipping and will control a false blow. The blade should meet the target at an angle of 45° for maximum effectiveness. A lesser angle will result in a glancing blow that may be dangerous to the user or other soldiers nearby. A greater angle will cause the blade to bounce, resulting in a minimum cut. All blows should be slanted away from the body to the left or right. Vertical blows should never be made.

d. Sharpening. The machete blade should be sharpened whenever necessary (*f*(2) below). Its edge should be sharpened from a point about 4 inches from the handle to the tip of the blade. The first 4 inches of the blade nearest the handle should not be sharpened.

e. Care. The machete handle should be kept smooth. It should not be wrapped with adhesive tape or cord nor should grooves or notches be cut in it. Such practices will cause the handle to blister the hands. All broken or cracked handles should be repaired or replaced. Care should be

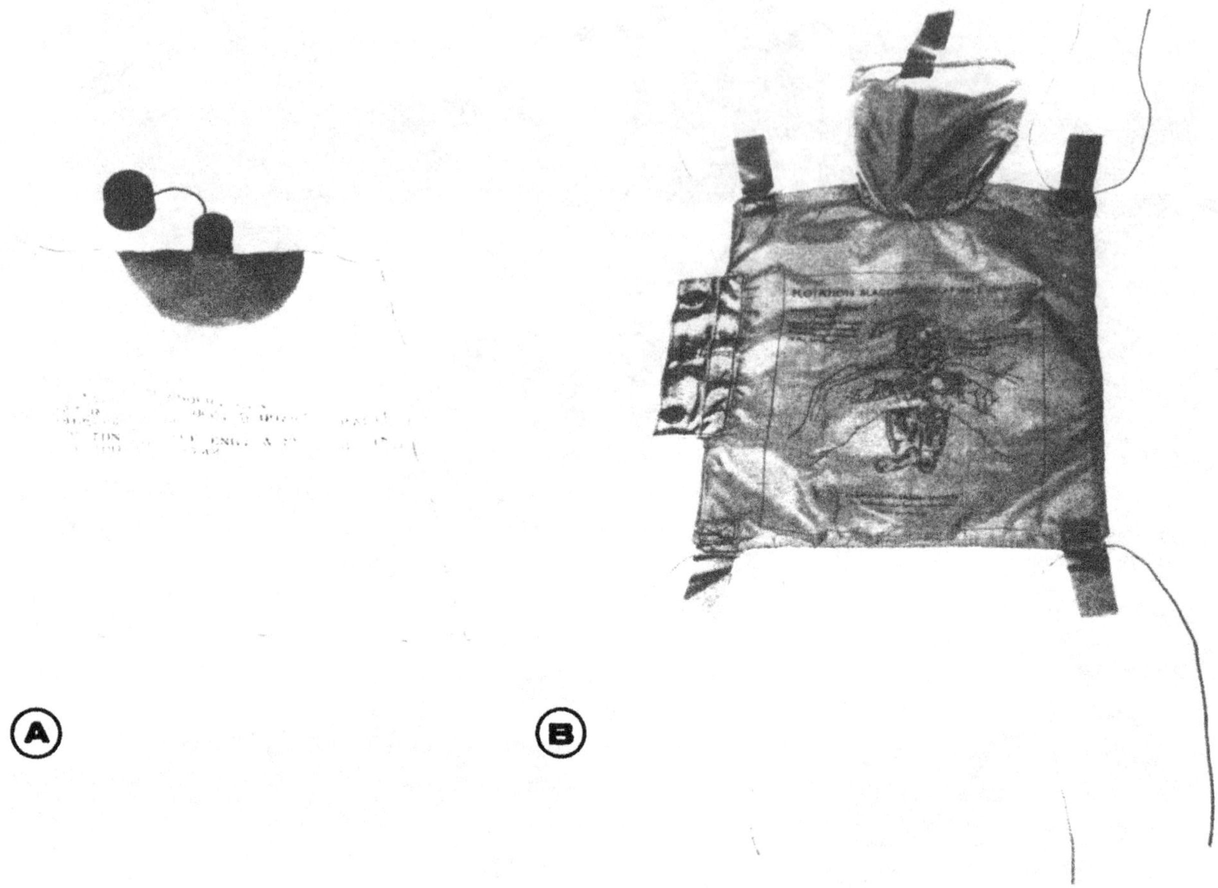

Figure 31. Flotation bladder assembly.

taken to avoid striking the machete against the ground. Striking the blade in the dirt or into stone objects will dull the cutting edge. The blade should be kept oiled or greased to prevent rust.

f. Sheath.

(1) *Description.* The plastic olive-green machete sheath (fig. 33) is 18¾ inches long and has a built-in tungsten carbide blade sharpener. A hook plate assembly is riveted at the top and is used to attach the sheath to the individual equipment belt.

(2) *Use and sharpening.* The machete sheath is used as a protective cover and carrying case for the M-1942 machete. To use the built-in sharpener:

(*a*) With the machete in the sheath, remove the individual equipment belt from the waist or remove the sheath from the belt.

(*b*) Rotate the sharpener so that the V-sharped notch of the sharpener projects slightly into the sheath opening.

(*c*) Place the edge of the sharpener against any flat surface.

(*d*) KEEPING FINGERS AWAY FROM THE SHARPENER, press down on the top of the sheath with one hand and pull the machete from the sheath with the other hand.

(*e*) Reinsert the blade for a second withdrawal *only* after pivoting the sharpener away from the sheath opening. One or two withdrawals should be sufficient for most sharpenings.

CAUTION

Do *not* hold sharpener in place by hand, either on the belt or off. This incorrect procedure may result in cutting off a finger.

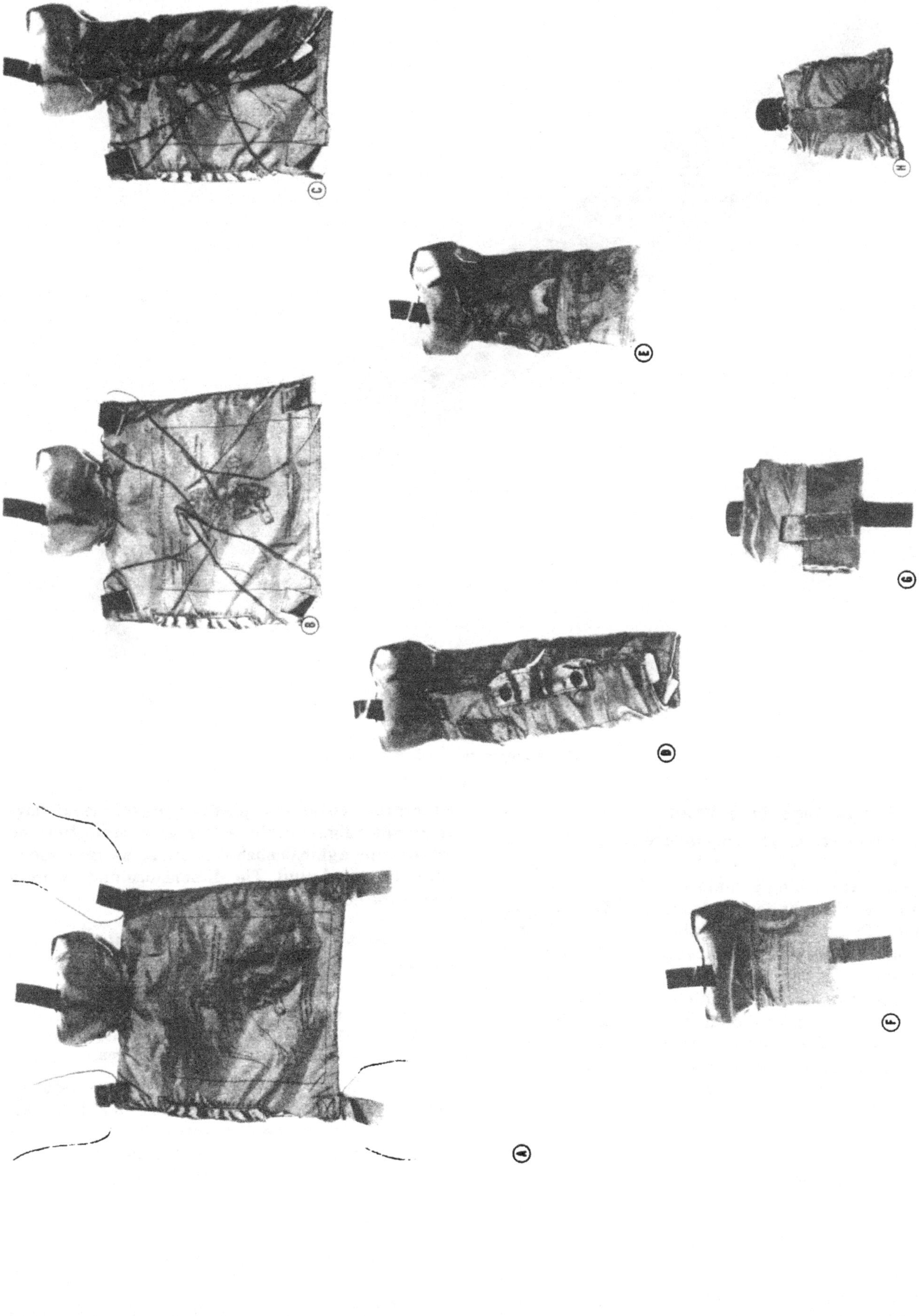

Figure 32. Folding flotation bladder assembly.

Figure 33. Machete with plastic sheath.

35. Camouflage Face Paint

a. Description. The camouflage face paint container consists of two cylinders joined end to end to form a single container, 3/4-inch in diameter. Each cylinder contains a different color paint. For hot weather the colors are light green and loam for tropic use and light green and sand for desert use. The paint contains an efficient insert repellent (diethyltoluamide). It is relatively odorless for security purposes. The consistency of the paint is not affected by aging.

b. Use. Camouflage face paint is used to cover or partially cover exposed skin areas to aid in concealment from the enemy (fig. 34).

36. Waterproof Covers

a. Description. The olive-green multipurpose waterproof cover is a plastic (polyethylene) bag. It provides lightweight, waterproof protection for small arms against such conditions as rain, moisture, dust, and mud. The dimensions and primary use of the covers are shown in table 3.

Table 3. Size and Use of Waterproof Covers

Size designation	Dimensions (inches)	Use
1	8 x 18	For pistol or personal effects.
2	10 x 56	For rifle or carbine.
3	15 x 45	For submachinegun.
4	20 x 84	For machinegun.

b. Use. When the waterproof cover is used for weapons, place the weapon in the cover, but first; twist and tie off the end with a simple knot. If the weapon is to serve as a flotation aid, partially inflate the cover before twisting and tying

Figure 34. Using camouflage face paint.

off the end to trap air in the cover and increase buoyancy. To remove the cover quickly, jerk the knotted end. In emergencies, the weapon can be fired through the cover. Other expedient uses of the waterproof cover are as follows:

(1) Use it to carry water.

(2) Use it to protect miscellaneous small items of food, clothing, and equipment.

(3) Inflate it for improvised lightweight watercrossing operations.

(4) Use it as a swimming aid by partially inflating it, tying off the end, and placing it inside the coat.

(5) Use it as an emergency waterproof clothing bag by using size 3 as a liner for the duffelbag. Cut approximately 21 inches off the top of the waterproof cover to make it a suitable length before inserting it into the duffelbag.

37. Poncho Shelters

Figure 36 illustrates various poncho shelters.

38. Helmet Band

The olive-drab helmet band is made of elastic cotton webbing. The band is worn around the outside of the helmet and is used to support camouflage materials.

39. Helmet Cover

The helmet cover is camouflage patterned and reversible. The cover is worn over the helmet for purposes of camouflage.

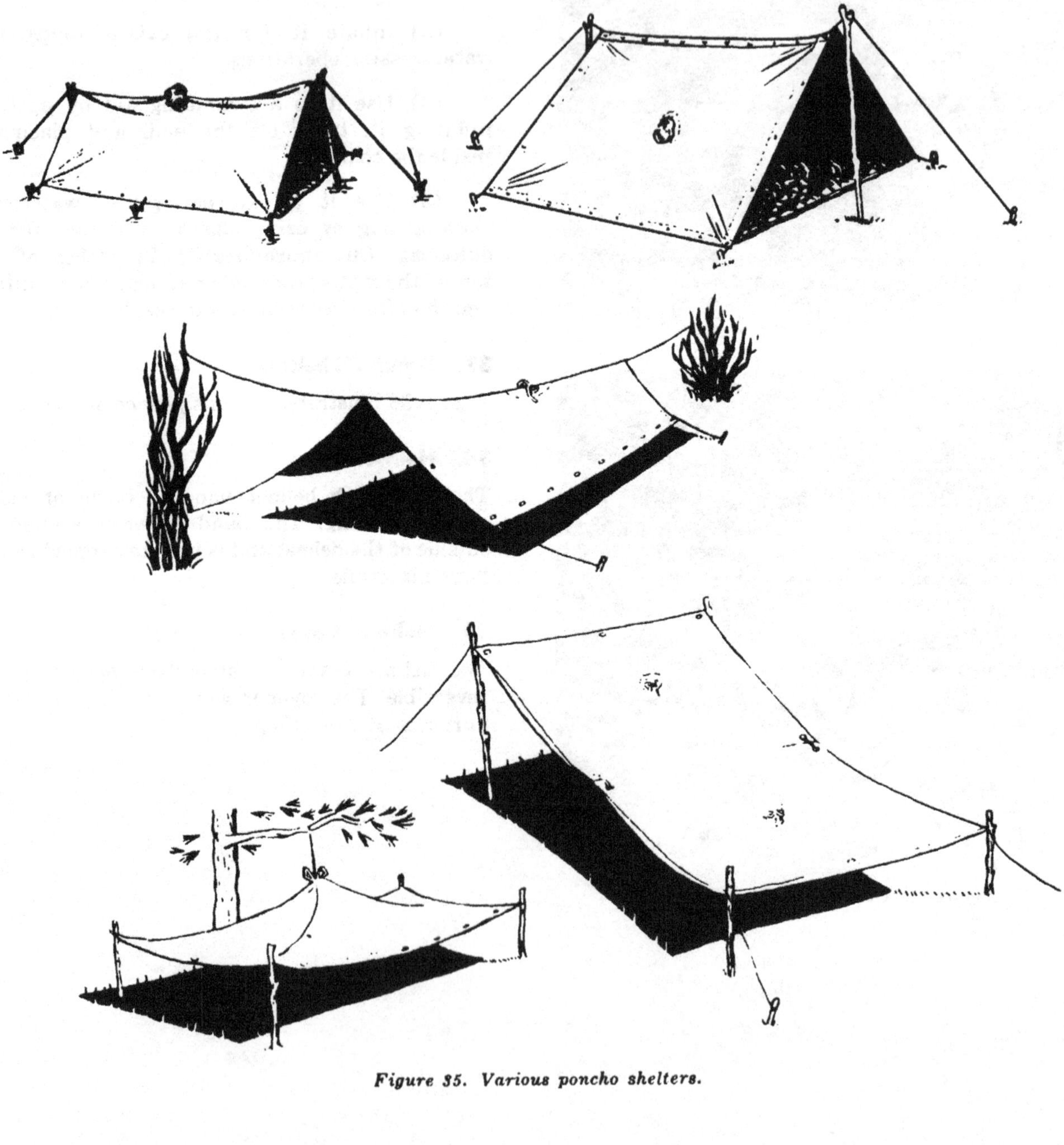

Figure 35. Various poncho shelters.

CHAPTER 6

CLEANING AND CARE

40. Individual Cleaning

a. General. When washing items of clothing and individual equipment, use a soap such as soft, hard, and sea water toilet soap or a detergent such as powdered laundry detergent. Brush or shake off any loose dirt or mud before washing the item. Thoroughly rinse all items, because soap or detergent residue left in the item may cause skin irritation and reduce the water repellency of the treated items.

b. Instructions.

(1) Wash underwear and tropical combat coat and trousers with soap or detergent and water, rinse thoroughly, *wring out*, pull garments into shape, and dry.

(2) Wash cushion sole socks, detachable tropical hat net, and canteen cover with soap or detergent and water, rinse thoroughly, *squeeze out* excess water, pull into shape, and allow to dry in the air.

(3) Wash tropical hat by hand only. Sponge or wash it gently with soap or detergent and water, being careful not to distort the shape. Rinse out all traces of soap or detergent, and allow to dry in the air.

(4) Wash the poncho liner and nylon knit sleeping shirt with soap or detergent and water, rinse thoroughly, and allow to *drip dry* in the air. Do not use hot water when washing these garments or hot air when drying them. *Do not* dryclean the poncho liner; drycleaning damages the bonded material.

(5) Brush or scrape off dried or caked dirt from the uppers and soles of the hot weather tropical boots, remove the laces and ventilating insoles. Wash the boots, laces, and insoles with soap or detergent and water. Wipe the insides of the boots with a clean cloth. Hang each boot (inverted) over a pole or stake to dry. *Do not* dry by exposure to a fire or by contact with a hot surface. Squeeze water out of laces and hang to dry. Allow insoles to dry before reinserting them into the boots.

(6) Wipe soiled coated items, such as the poncho and pneumatic mattress, with a clean cloth, wash with water and a mild soap or detergent, and rinse thoroughly. Dry the items as quickly as possible, but do not use direct heat as this causes coated material to crack. Take special care to see that such substances as oil, grease, acid, or insect repellent are washed off as soon as possible.

41. Unit Laundry Cleaning

Refer to TM 10-354 or TM 10-280 for proper washing and drying procedures. The tropical hat should *not* be machine laundered.

42. Individual Care

a. Body Clothing. Repair holes and tears with quick stick-on patches (fig. 36). The patches enable clothing to be repaired quickly, even while being worn (fig. 37). When clothing is wet, the area to be patched should be dried first.

b. Head Net. Repair small tears and holes by placing pieces of adhesive tape or waterproof tape over both sides of each hole while net is flattened out. Edges of small holes may also be drawn together with a needle and fine thread. Use a small rubber band or nylon cord as an emergency replacement of the elastic loops.

c. Multipurpose Net. Lace broken edges of the net together, using small nylon cord.

d. Pneumatic Mattress.

(1) Do *not* inflate the pneumatic mattress with air lines or other mechanical means of inflation. Excessive pressure may tear the cemented seams. Take care not to place the mattress on sharp objects that may puncture it. To detect small holes, dip the inflated mattress in water and look for air bubbles.

(2) Repair small holes and tears by following the instructions in the pneumatic mattress repair kit.

(3) When packing the mattress, always roll it toward the open valve to release all the air.

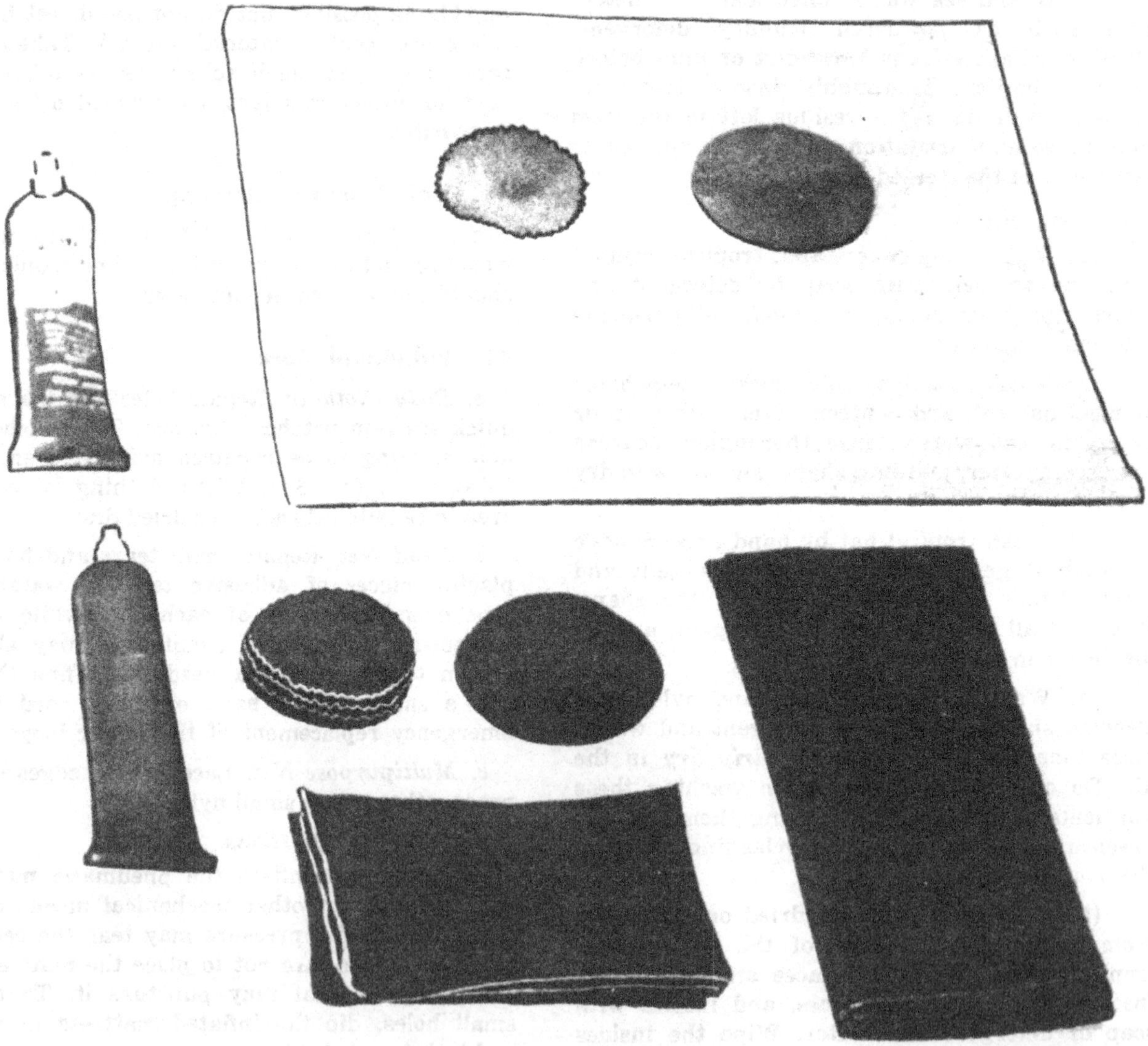

Figure 36. Quick stick-on patch kit.

Place the mattress on the sleeping equipment and roll together, or roll the mattress separately and put it in or attach it to the field pack.

e. Coated Items. When storing coated items, make certain the items are dry before folding them. When possible, fold so that a coated side will be against an uncoated side to avoid sticking. When packing coated items make certain that they do not come in contact with sharp objects that may cause rips or tears. When possible, dust coated items with talcum before storing.

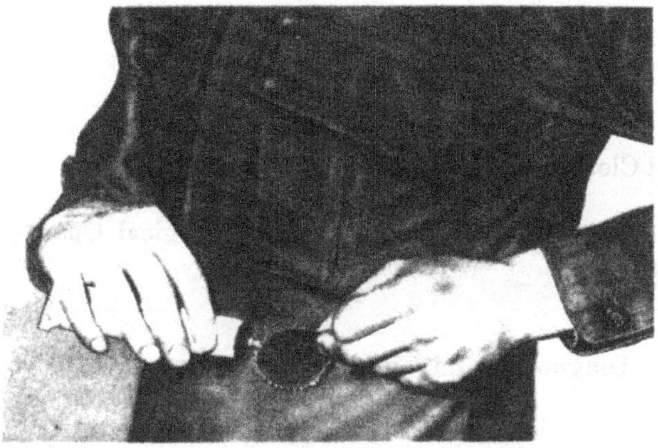

Figure 37. Applying quick stick-on patch while wearing garment.

APPENDIX A

REFERENCES

DA Pam 95-6	Hot Weather Sense
DA Pam 95-10	Survival Sense
FM 21-10	Military Sanitation
FM 21-11	First Aid for Soldiers
FM 21-13	The Soldiers' Guide
FM 21-15	Care and Use of Individual Clothing and Equipment
FM 21-18	Foot Marches
FM 21-41	Soldier's Handbook for Defense Against Chemical and Biological Operations and Nuclear Warfare
FM 21-76	Survival
FM 31-72	Mountain Operations
TB MED 175	The Etiology, Prevention, Diagnosis, and Treatment of Adverse Effects of Heat
TM 10-227	Fitting of Men's Uniforms
TM 10-228	Fitting of Footwear
TM 10-267	General Repair for Clothing and Textiles
TM 10-268	General Repair of Footwear and Leather Goods
TM 10-269	General Repair for Canvas and Webbing
TM 10-270	General Repair of Quartermaster Items of General Equipment
TM 10-277	Protective Clothing—Chemical Operations
TM 10-280	Field Laundry, Bath, and Clothing Exchange Operations
TM 10-354	Army Fixed Laundry Organization
TC 10-8	The Lightweight Rucksack: Nylon OG 106 (With Riveted Frame)
SB 10-523	Size Tariff for Clothing, Equipage, and Footwear
SB 700-20	Army Adopted Items of Materiel and List of Reportable Items
ASubjScd 10-30	Global Use of Field Combat Clothing and Individual Equipment
TF 3-4060	Components of the M17A1 Field Protective Mask
TF 3-4061	Accessories for the M17A1 Field Protective Mask
TF 10-2426	Fitting Army Uniforms and Footwear—Fitting of the Army Green Uniform
TF 10-2430	Fitting Army Uniforms and Footwear—Fitting of Footwear
TF 10-3092	Individual Load Carrying Equipment
TF 10-3593	Individual Load Carrying System—Variations of Use
TF 10-3752	Lightweight Rucksack
T (GTA) 10-4-3	Care of Quartermaster Clothing and Equipment
T (GTA) 10-4-6	Fragmentation Protective Body Armor (Composite Vest)
GTA 10-4-8	Fitting of Clothing
GTA 10-35	Use and Care of Footwear

APPENDIX B

SHORT FEDERAL STOCK NUMBER NOMENCLATURE

The following table lists the short Federal stock number nomenclature* for identification purposes only. The items are grouped by clothing and by equipment and are listed alphabetically within those groups with text paragraph references and figure references, when appropriate.

Nomenclature	FSN	Text and figure references
CLOTHING		
BOOTS HOT WEATHER: Tropical men's lthr and nyl duck DMS spike resistant (panama sole).	8430-141-0767-series	para 17, fig. 9.
CAP HOT WEATHER: Polyester and rayon olive green 106.	8415-177-4834-series	para 11, fig. 2.
COAT CAMOUFLAGE: Printed camouflage pattern	8415-945-7650-series	para 9.
COAT HOT WEATHER: Cotton wind resistant poplin	8405-935-4702-series	para 9, fig. 1.
DRAWERS MENS: Cotton thigh length OG 109 w/elastic waistband.	8420-782-6405-series	para 8.
DRAWERS MENS: Cotton thigh length white w/elastic waistband.	8420-682-6593-series	para 8.
HAT AND INSECT NET: Camouflage pattern AG 323 w/chin strap band.	8415-141-0819-series	para 12, B, fig. 3.
HAT AND INSECT NET: With chin strap olive green 107	8415-935-2885-series	para 12, A, fig. 3.
NECKERCHIEF MAN'S: Ctn knit OG 109 f/tropical combat.	8440-935-6374	para 13, fig. 4.
PONCHO: Coated nylon olive-green shade 207	8405-935-325	para 14, fig. 5.
PONCHO: Nylon lightweight camouflage printed pattern:		para 14.
Regular	8405-935-6621	
Small	8405-935-6859	
SHIRT SLEEPING HEAT RETENTIVE AND MOISTURE RESISTANT: Pullover.	8415-890-2100-series	para 25, fig. 19.
SOCKS MEN'S: Wool cushion sole OG 408 stretch type	8440-782-2171-series	para 16.
TROUSERS CAMOUFLAGE Printed camouflage pattern	8415-945-9212-series	para 10.
TROUSERS MEN'S: Cotton wind-resistant poplin	8405-935-3302-series	para 10, fig. 1.
UNDERSHIRT MANS: Cotton OG 109 pullover qtr length sleeves.	8420-782-6707-series	para 8.
UNDERSHIRT MANS: Cotton white pullover quarter length sleeves.	8420-543-6643-series	para 8.
EQUIPMENT		
BAND HELMENT CAMOUFLAGE: Mildew-resistant water-repellent OD.	8415-576-2873	para 38.
BELT INDIVIDUAL EQUIPMENT: Nylon webbing OD Army shade 7:		para 28, fig. 22.
Medium	8465-935-6815	
Large	8465-935-6816	
BLADDER ASSEMBLY FLOTATION: Vinyl plastic 5-qt collapsible.	8465-141-0924	para 33, fig. 31.
CANTEEN WATER: Collapsible	8465-927-7484	para 32, fig. 29.
CANTEEN WATER: Plastic	8465-889-3744	para 32, fig. 28.
CARRIER INTRENCHING TOOL: Webbing olive-drab MRWR.	8465-542-5842	para 28, fig. 22.
CARRIER SLEEPING EQUIPMENT: OG 106	8465-935-6813	para 28, fig. 22.

*SB 700-20, Army Adopted Items of Materiel and List of Reportable Items.

Nomenclature	FSN	Text and figure references
CASE FIELD: First aid dressing unmounted magnetic compass OD cotton.	8465-577-4927	para 28, fig. 22.
CASE SMALL ARMS AMMUNITION: OD 7 M16 rifle 4-20 rounds mag.	8465-935-4871	para 28, fig. 22.
COVER HELMET CAMOUFLAGE PATTERN: Reversible.	8415-261-6833	para 39.
COVER WATER CANTEEN: Webbing OD 7 MRWR	8465-577-4926	para 28, fig. 22.
COVER WATER CANTEEN: Collapsible	8465-927-7485	para 28, fig. 22.
COVER: Waterproof machine gun 20 in. W 84 in. L	8465-185-0725	para 36.
COVER: Waterproof pistol 8 in. W 18 in. L	8465-185-0726	para 36.
COVER: Waterproof rifle or pistol 10 in. W 56 in. L	8465-185-0723	para 36.
COVER: Waterproof submachine gun 15 in. W 45 in. L	8465-185-0724	para 36.
FIELD PACK CANVAS: Combat	8465-577-4921	para 28, fig. 22.
HAMMOCK JUNGLE: Nylon with cord	8465-935-6397	para 22, fig. 10.
LINER WET WEATHER PONCHO: Camouflage color	8405-889-3683	para 23, fig. 12.
MACHETE: M-1942	5110-813-1286	para 34, fig. 33.
MATTRESS PNEUMATIC: Nylon coated two sides OG 207 7¾ x 31½ in.	8465-254-8887	para 25, fig. 17.
NET MULTIPURPOSE: Nylon 108 in. L 60 in. W	8465-889-3771	paras 24, 29, and 31, figs. 26 and 27.
PAINT FACE: Camouflage:		para 35, fig. 34.
Light green and loam	8510-161-6204	
Light green and sand	8510-161-6202	
RUCKSACK: Nylon OG riveted frame	8465-782-3248	para 27, fig. 21.
RUCKSACK: Tropical lt wt nylon w/waterproof liners	8465-935-6673	para 26, fig. 20.
SHEATH MACHETE: Plastic OD w/built-in sharpener f/machete 18 in. L 2¼ in. W.	8465-926-4932	para 34, fig. 33.
SUSPENDERS FIELD PACK: H-type:		para 28, fig. 22.
Regular cotton	8465-577-4922	
Long cotton	8465-577-4923	
Extra long cotton	8465-823-7231	
VEST GRENADE CARRYING: Nylon duck OG 106 (ammo f/M-79 launcher).		para 30, fig. 25.
Small	8415-141-0926	
Large	8415-141-0927	

By Order of the Secretary of the Army:

W. C. WESTMORELAND,
General, United States Army,
Chief of Staff.

Official:
KENNETH G. WICKHAM,
Major General, United States Army,
The Adjutant General.

Distribution:

To be distributed in accordance with DA Form 12-11 requirements for Desert Operations and Jungle Operations.

☆ U.S. GOVERNMENT PRINTING OFFICE: 1970—431-133/5020

Made in the USA
Monee, IL
08 November 2026

67994719R00031